Accession no.
36076408

RESEAR
IN HIGHER EDUCATIO

D0303912

Across the world universities are transforming their teaching and learning practices to meet the challenges facing Higher Education in the 21st century. The best of this work is based on high-quality research and excellent scholarship. This much-needed book provides a basic and comprehensive approach into the research methods which are now significantly improving teaching and learning practices in many countries.

A complete resource for lecturers of any discipline, professional developers, researchers and graduate students, this book covers the essential methodological and theoretical foundations needed to engage in Higher Education research. The book has a refreshingly light yet serious tone, and takes a deeply humane approach to researching the human relationships between student and teacher.

Each chapter combines an introduction to relevant theoretical concerns with "how to" guides and resources for further study.

Key themes are:

- Epistemological and ethical frameworks
- Qualitative data analysis
- Focus group, semi-structured interviews and narrative inquiry
- Case study research and ethnographic approaches
- Action research and appreciative inquiry
- Phenomenographic approaches and visual research methods
- Researching threshold concepts
- Evaluation research

This book will be an invaluable resource for anyone interested in the up-to-date theories and methods for conducting teaching and learning research in Higher Education.

Glynis Cousin is Professor of Higher Education and Director of the Institute for Learning Enhancement at the University of Wolverhampton. She is one of the UK's foremost higher educational researchers and has extensive experience working with teachers in many disciplines. She has written widely on education research and evaluation, curriculum inquiry, diversity and internationalization of the curriculum.

The Staff and Educational Development Series

Series Editor: James Wisdom

SEDA is the Staff and Educational Development Association. It supports and encourages developments in teaching and learning in higher education through a variety of methods: publications, conferences, networking, journals, regional meetings and research – and through various SEDA Accreditation Schemes.

SEDA
Selly Wick House, 59–61 Selly Wick Road, Selly Park, Birmingham B29 7JE
Tel: 0121 415 6801
Fax: 0121 415 6802
E-mail: offlce@seda.ac.uk
Website: wwwsedaac.uk

RESEARCHING LEARNING IN HIGHER EDUCATION

An Introduction to Contemporary
Methods and Approaches

Glynis Cousin

LIS LIBRARY

Date	Fund
20/07/09	Iti

Order No

2024275

University of Chester

Routledge
Taylor & Francis Group

NEW YORK AND LONDON

First published 2009
by Routledge
270 Madison Ave, New York, NY 10016

Simultaneously published in the UK
by Routledge
2 Park Square, Milton Park, Abingdon, Oxon OX14 4RN

Routledge is an imprint of the Taylor & Francis Group, an informa business

© 2009 Taylor and Francis

Typeset in Adobe Caslon and Trade Gothic by
Florence Production Ltd, Stoodleigh, Devon
Printed and bound in the United States of America
on acid-free paper by Sheridan Books, Inc.

All rights reserved. No part of this book may be reprinted or
reproduced or utilized in any form or by any electronic,
mechanical, or other means, now known or hereafter invented,
including photocopying and recording, or in any information
storage or retrieval system, without permission in writing from
the publishers.

Trademark Notice: Product or corporate names may be
trademarks or registered trademarks, and are used only for
identification and explanation without intent to infringe.

Library of Congress Cataloging in Publication Data
    Cousin, Glynis.
        Researching learning in higher education:
an introduction to contemporary methods and approaches/
Glynis Cousin.
            p. cm.
        Includes bibliographical references and index.
    1. Education, Higher—Research—Methodology.
    I. Title.
    LB2326.3.C685 2009
    378.0072—dc22   2008027784

ISBN10: 0–415–99164–1 (hbk)
ISBN10: 0–415–99165–X (pbk)
ISBN10: 0–203–88458–2 (ebk)

ISBN13: 978–0–415–99164–3 (hbk)
ISBN13: 978–0–415–99165–0 (pbk)
ISBN13: 978–0–203–88458–4 (ebk)

For Shoshi and Phindi

Contents

CONTENTS

Acknowledgments

I am deeply indebted to James Wisdom, the series editor for SEDA, for the huge support he has given throughout the writing of this book. Many thanks also to Sarah Burrows, Routledge, for her warm encouragement and support. A number of people have helped with particular chapters—Liz Beaty, Ray Land, Erik Meyer, Mick Healey, Russell Goodwin, Brendan Hall, Martin Jenkins, Laura Lannin, Jo Lonsdale, Wendelin Romer and Sue Swansborough. Many thanks also to those who kindly sent their work and references: Paul Ashwin, Tamsin Haggis, Marina Orsini-Jones, Jane Osmond and Alison Shreeve. Many thanks to Bill Dunn for his data on students' stories. I am also very grateful to Maggie Savin Baden who generously shared her data analysis strategy for narrative inquiry. Finally, thanks to Robert Fine for his much appreciated support and to Shoshi Fine who allowed me to use her interview data.

Foreword

This is a book we have needed for some time, and I am delighted that Glynis Cousin has written it. She has brought to it a deep respect for the quality of human relationships, and has elevated pedagogic research into a noble practice which engages the heart as well as the head.

Perhaps one of the most exciting features of world-wide higher education at the moment is the growth of a genuine enthusiasm in those who teach (and those who help students to learn) to research, understand and improve the experience of learning. All of these people—be they faculty, academics, lecturers, staff or teachers—have been steeped in their discipline, they have become skilled practitioners of its core processes and are committed to its success. They may find researching into learning, studying their students and working with pedagogic evidence a small step or a long journey. I hope they will come to recognize the value of this book as their guide.

One of the most interesting features in the process of commissioning and supporting this book lay in how it exposed the view that there is "proper" research into higher education, conducted by practitioners who have committed to this specific discipline, creating hard-edged data so reliable that governments could plan policy on it. We are undoubtedly witnessing the creation of a new academic subject—the pedagogy of higher education—and those who lead this work deserve our gratitude and respect. However, in one significant dimension this new subject may not resemble so many others, and that difference lies in the concept of "the dual professional." The historian, the chemist, the land surveyor, the nurse—each devoted to their subject, each committed heart and soul to their students. It is for them that this book has been written.

James Wisdom
SEDA Series Editor
Visiting Professor in Educational Development,
Middlesex University, UK

1

FRAMING HIGHER EDUCATION RESEARCH

Researching Teaching and Learning in Higher Education

The aim of this book is to provide a sufficient basis for each method covered to support readers to use it straightaway. Because methods need to be underpinned by theory, I have given equal weight in each chapter to theoretical concerns as well as to practical steps. Methods are never neutral tools and good research requires an engagement with their theoretical underpinning.

Needless to say, I do not claim to exhaust all the approaches that can be used for higher education research. Those that I have selected simply reflect my own experience as a sociologist, teacher and researcher in higher education. All of these approaches are qualitative though, as I explain below, I do not subscribe to a simplistic qualitative/quantitative opposition.

The chapters on ethics and data analysis are located at the beginning of this book to encourage the reader to address these issues alongside whatever approach they choose. Most of the methods I discuss require the researcher to gather and analyze as contemporaneous rather than sequential activities. Similarly, most of the approaches require a sustained attention to the ethical dimension throughout the research cycle. I urge readers to look at these chapters first.

I do make overlapping points and repeat advice across the chapters but this is deliberate in order to limit cross-referencing. I think it is irritating for the reader to have to constantly shuffle back and forth between chapters. Moreover, where I do repeat points and advice, I try to provide a fresh angle on them. Shot through this book is a value position which is based on the following cautions.

Method as Servant

Firstly, do not let the tail wag the dog—research methods are in the service of the researcher, not vice versa. Treat rules about methods as guidelines which you can adapt, refine, expand or trim. I am not suggesting here that anything goes. As Coffey and Atkinson (1996: 11) rightly object, some practitioners "believe that qualitative research can be done in a spirit of careless rapture." All research, quantitative or qualitative, has a strong craft dimension to it and this involves knowing some of the rules before throwing them away.

An Age of Emancipation

Secondly, heed the words of Denzin and Lincoln (2000: 162):

> This is an age of emancipation; we have been freed from the confines of a single regime of truth and from the habit of seeing the world in one color.

While randomized control trials remain a gold standard for some researchers, particularly in the field of medical research, education studies has become a big playground where no one methodology needs to hog the best swing. The good researcher knows how to play around with many possible approaches in a spirit of curiosity about what they can yield. In this "age of emancipation," this spirit promises to replace the immaturity of paradigm warfare.

Good Thinking

Thirdly, heed the words of Stake (1995: 19):

> Good research is not about good methods as much as it is about good thinking.

This is a caution against a narrow view of empirical research in which the hunting and gathering of data are treated as the primary purpose of the research enterprise. Equally important is the ongoing scholarship you bring to the research cycle so that your work has a strong intellectual direction. Much of the time, you need to be thinking *with* the data as much as you are thinking *from* it (Cousin, 2007). As indicated, many contemporary approaches require that you move away from the conventional linear processes of firstly, a literature review, secondly question formulation, thirdly data collection, fourthly data analysis and fifthly the write-up. Increasing numbers of researchers recognize that all of these activities need to be dynamically linked and continually enlivened by an engagement with a wide reading.

Understanding Traditions

Fourthly, avoid deference to particular paradigms and do not get overly involved in paradigm wars. Bear in mind that paradigm warfare always carries with it the heavy risk of otherizing the apparent adversary. Whatever framework we choose, the overarching one has to be that of respect for different cultures of inquiry (to use Bentz and Shapiro's (1998) helpful term) unless there are some very clear reasons to withhold that respect. Moreover, this needs to be informed respect in which we take the trouble to open some of the books which may not seem to our taste in order to avoid stereotypification of them.

If, like myself, you have a leaning towards the qualitative, have a look at Stephen Gorard's (2006) very useful, slim volume *Using Everyday Numbers Effectively in Research* and this will increase your respect and understanding for the quantitative dimension of any research. Incidentally, this book shows how numbers can be used in research by even the most innumerate of us. Similarly, if you are sceptical of the "scientific" paradigm, browse through the accessible, thoughtful chapters in the hefty *Handbook of Research Design in Mathematics and Science Education* (Kelly and Lesh, 2000) and you will discover commitments to uncertainty, fuzziness, humor and interpretation which are seriously under-recognized by opponents to the scientific tradition.

If you are interested in survey research into student learning experiences, Edinburgh University's project *Enhancing Teaching and Learning in Undergraduate Education Project* has a helpful set of instruments www.tla.ed.ac.uk/etl/publications.html#measurement.

If you have a preference for the quantitative, take a look at Robert Stake's (1995) *The Art of Case Study Research*, or perhaps Eisner's (1991) *The Enlightened Eye*, or maybe just read a classic ethnographic text and appendix such as Foote-Whyte's (1993) *Street Corner Society*. In all of these books, you will discover a commitment to careful method and to the provision of trustworthy, analytical, evidence-informed accounts that should impress any statistician's need for detail and validity.

Having compared the quantitative with the qualitative, I do not want to suggest that these are distinctive paradigms (a frequent misunderstanding) since there are many approaches and internal contestations within and across quantitative and qualitative frameworks. Moreover, the two approaches overlap more than is commonly recognized. They are not strict opposites. Both quantitative and qualitative research involves interpretation, as Stake wrote (1995: 9):

> Interpretation is a major part of all research. I am ready to argue when someone claims there is more interpretation in qualitative research than in quantitative.

While there may be different degrees of interpretation brought to bear on different sets of figures, there is always some form of interpretation going on. At the very least, the quantitative researcher has to make a decision about which figures to gather, which variables to select, which associations or correlations to make and what to frame as noteworthy for the reader of the analysis. All of these are acts of interpretation. And inferential statistics are called thus because a lot of inferring is involved in the analysis.

Similarly, much qualitative research involves counting of some kind. Notions (discussed in various chapters) of theoretical sampling and saturation from grounded theory (Glaser and Strauss, 1967) have a quantitative logic. Similarly, content analysis is a quantitative move within the qualitative tradition. More generally, qualitative researchers attend to frequencies, absences and patterns in their data. Indeed, it is hard to look at text-based qualitative data without searching for some kind of pattern in it though it is the novice's error to do so at the cost of noting singularities and difference.

Some research questions require a quantitative orientation, including sophisticated statistical analysis; some a qualitative orientation; and some an epistemologically aware mix of the two. It all depends on the problem in hand. If I want to know if an increase in female take-up of engineering courses is connected to an expanded school curriculum, I will want to do some quantitative research. If I want to know what it is like to be a minority of males studying midwifery, I will probably want to conduct qualitative research. Scorn for either quantitative or qualitative sources of intelligence is unintelligent, as is the claim that one source of intelligence is superior and more reliable than the other. One notion that needs defeating is that of qualitative research as the ornament of quantitative research; and another is that of quantitative research as hopelessly objectivist.

Treading a Middle Path

Fifthly, by urging respect for the various perspectives, I am not suggesting that you adopt a mindless eclecticism that takes no account of methodological debates. Rather, I am sharing Silverman's (1997: 1) desire "to search for ways of building links between social science traditions rather than dwelling in 'armed camps' fighting internal battles." I am also heeding Oakley's (1999: 155) words:

> The main point about paradigms is they are *normative*; they are ways of breaking down the complexity of the real world that tell their adherents what to do . . . a way of life rather than a set of technical and procedural differences.

To support the reader to "build links" between traditions in the way Silverman suggests, I will sketch briefly the methodological issues of particular relevance to the approaches presented in this book.

Method and Methodology

Although many methods are friendly to particular methodologies and to particular research contexts or questions, there is not always a straightforward association between method and methodology. This is because different people might use the same methods with quite different values and aims in mind.

Crudely, methods are best understood as the tools and procedures we use for our inquiries and methodology is about the framework within which they sit. For instance, experimental design usually requires adherence to a set of procedural methods that have to do with the careful selection of subjects, the setting, the timing and implementation of the experiment, the logging of findings and their analysis. Methodologically these procedures—which aim to measure effects and to eliminate researcher bias or interference—are defended on the grounds that, if faithfully replicated, they will yield the same findings. Validity will be seen to depend on the accuracy of the testing instrument, the comparability of the subjects participating and the absence of "contaminating" factors that might disturb the results. A careful deployment of the methods will produce a methodologically sound picture of what is going on.

In the case of much contemporary ethnography, there is a set of quite different but equally careful methods concerning research access, observations, interviewing, record keeping and data analysis. Taken together, these methods feed into a methodological framework which expects the findings, analysis and write-up to be generative of further understandings about the setting under investigation. I will return to this contrast later.

Epistemology and Ontology

A connected question to the method/methodology twosome concerns questions of epistemology and ontology. All researchers have to address these questions at some level because, as Usher et al. (1997: 176) write, every research method is "embedded in commitments to particular versions of the world (ontology) and ways of knowing that world" (epistemology).

Epistemology is about conceptions of the nature of knowledge and of ways of coming to know and ontology is about conceptions of our positionality in the world and the effects this has on what is knowable. In other words, what can we see from where we stand in relation to the research setting? Are we inside the research setting, at a distance from it or somewhere in between? Do we have a transparent view on to reality or will it always be mediated by our subjectivity? Our responses to these questions frame our epistemological and ontological stance.

In order to draw out some of the epistemological and ontological contrasts across research traditions, let me illustrate with two report extracts—the first concerns experimental design research into a US schooling scheme, Even Start. This public scheme publishes its ongoing evaluation results on its website:

> **Findings:** As was the case at posttest, Even Start children and parents made gains on a variety of literacy assessments and other measures at follow-up, but they did not gain more than children and parents in the control group. It had been hypothesized that follow-up data might show positive effects because (1) Even Start families had the opportunity to participate for a second school year, and (2) change in some outcomes might require more time than others. However, the follow-up data do not support either of these hypotheses. **Conclusion:** The underlying premise of Even Start as described by the statute and implemented in the field was not supported by this study. (Even Start, 2008: 7)

In this example, the researchers have discovered through careful, objective, testing that Even Start does not make any difference to children's school performance. The research design is likely to be termed positivist because it seems to be committed to the belief in a knowable world that is separable from the researcher. Associated with this is a commitment to the possibility of identifying underlying regularities within this world that will enable predictions to be made.

Now a very different kind of research report from a famous piece of ethnography from Clifford Geertz (1973: 413):

> On the established anthropological principle, When in Rome, my wife and I decided, only slightly less instantaneously than everyone else, that the thing to do was run too. We ran down the main village street, northward, away from where we were living, for we were on that side of the ring. About half-way down another fugitive ducked suddenly into a compound—his own, it turned out—and we, seeing nothing ahead of us but rice fields, open country, and a very high volcano, followed him. As the three of us came tumbling into the courtyard, his

wife, who had apparently been through this sort of thing before, whipped out a table, a tablecloth, three chairs, and three cups of tea, and we all, without any explicit communication whatsoever, sat down, commenced to sip tea, and sought to compose ourselves.

This extract sits in a post-positivist or hermeneutic/interpretivist framework. Henceforth, I am going to use the umbrella term interpretivist for this tradition. I use this broad term to embrace any perspective that foregrounds the search for meanings (discussed below). Since the researchers in this framework judge objectivity to be impossible in the human sciences, they freely insert themselves in the research process and the report. They are part of the setting rather than outside of it. This tradition handles subjectivity in the research process by owning up to it. Further, researchers in this framework aim for some measure of closeness with subjects because rapport building and respondent disclosure are seen to be interdependent. They argue that people will not tell you what is happening unless they trust you and trust cannot be built by keeping one's distance.

Notions of validity are replaced by those of trustworthiness within interpretivism since all the approaches I deal with in this book are broadly from this tradition with some leaning more towards a post-modern concern for the determining nature of language than others (of which more below). I discuss issues of trustworthiness for each of the approaches but, broadly, it is commonly held to be secured through moves such as triangulation (comparing different data sources), and/or through checking accounts with research subjects, demonstrating researcher reflexivity, collecting and surfacing sufficient data for plausibility and providing rich descriptive and analytical accounts.

One of the enduring confusions among different research communities comes from many researchers evaluating interpretivist research against traditional validity criteria. Some find it hard to grasp the merit of going deep rather than wide, of putting the researcher in the picture and not having control groups. Similarly, interpretivists often overdraw the objectivism or commitment to certainty of traditional, scientific cultures of inquiry. Most researchers associated with these cultures are well aware of the impossibility of total objectivity and of the presence of error in their investigations, which is why they tend to talk of probability rather than certainty. Each tradition is best understood as having *ideals* which inform the direction of research.

There are three fundamental differences in orientation between the positivist ideals and those of interpretivism. Firstly, the latter concerns "a systematic empirical inquiry into meaning" (Shank, 2002: 11) and it is this interest in meaning that is held to set interpretivists apart from the positivist tradition. Interpretivists argue that the human sciences must address people's intentions within given contexts, not simply observe their outward behavior.

I may salivate at the sight of food but resist eating it; my resistance can only be understood by exploring the meanings I make of the situation. Perhaps I am on a diet. At any rate, an explanation of my resisting behavior is not forthcoming from a simple observation. So an inquiry into human meaning is never going to be as straightforward as observing the behavior of salivating dogs and this brings us to the second difference between the two broad traditions.

Cultures of inquiry associated with positivism could be said to pursue *explanations* of and *predictions* about human behavior, while those associated with interpretivism aspire to generate *understandings and insights* in contexts that are held to be inherently too unstable for reliable predictions to be made. Again, let me caution against overdrawing these distinctions because research reports from the former often end with a call for the need for further exploration and from the latter quite firm claims are often made, particularly those with a concern for power and the emancipatory function of research. The third area of difference concerns research report genres.

Writing Difference

A third contrast between positivist and interpretivist traditions concerns their respective use of language and the researcher positionality this expresses. Shank (2002: 10) nicely exemplifies this contrast by suggesting that whereas a positivist might write "these effects were observed," an interpretivist is more likely to write "I observed these effects." The two extracts above are good examples of the different writing genres in the two traditions. Foley (1998: 110) talks of the "scientific realism" evident in the first phrase:

to evoke an authoritative voice, the author must speak in the third person and be physically, psychologically, and ideologically absent from the text. That lends the text an aura of omniscience. The all-knowing interpretive voice speaks from a distant, privileged vantage point in a detached, measured tone.

To avoid what they see to be feigned objectivity of this kind, the interpretivist researcher tends to be out and proud with the first person, as Eisner (1991: 4) writes:

> I want readers to know that this author is a human being and not some disembodied abstraction who is depersonalized through linguistic conventions that hide his signature.

As we will see in a moment, the interpretivist acknowledgment that research report genres are culturally formed is extended by the postmodern, social constructionist concern for the determining power of language. This concern is characterized sometimes as the "linguistic turn" in social science.

Mindfulness

The adoption of the first person by qualitative researchers has fuelled a popular conception that because the interpretivist tradition accepts the impossibility of removing the subjective, it abandons any notion of objectivity. This is not quite the case, as Geerz (1973: 16) writes:

> I have never been impressed by the argument that, as complete objectivity is impossible in these matters (as, of course, it is), one might as well let one's sentiments run loose. As Robert Solow has remarked, that is like saying that as a perfectly aseptic environment is impossible, one might as well conduct surgery in a sewer.

A contemporary spin on this statement claims mindfulness rather than objectivity (Bentz and Shapiro, 1998) for the interpretivist tradition. This concerns a resistance to the "careless rapture" of which Coffey and Atkinson (1996) speak above and attention to the honesty and plausibility of research processes and accounts. But there is another twist to the language question that needs addressing.

As I have indicated, interpretivism is a very broad umbrella, uniting all perspectives that place an emphasis on human meaning-making activities (particularly, phenomenology, symbolic interactionism and constructivism). My use of the term in this book also embraces *some* of the insights of the postmodern perspective of social constructionism. Because of the current influence of social constructionist debates in qualitative research, I will outline what I think is needed to address this within the spirit of Seale's (1999: 31) position that:

> Research is a craft skill, relatively autonomous from the need to resolve philosophical or epistemological debates, but it can nevertheless draw on these as resources in developing methodological awareness.

Social Constructionism

Following Kuhn's (1962) pioneering inquiry into research as a cultural practice, social constructionists argue that research methods construct social realities as much as they might describe or "discover" them. One reason for this is that we can only *represent* reality, we can never mirror it and the act of representation is always going to be adrift from the event. Crudely put, my trip to the dentist and my narrative about it are always going to be two different things. Schostack (2002: 2) explains the problem:

> So here is the paradox: the lived always seeks to be represented in some way and thus sacrifices the sense of life for the sense of words and meanings in order to relive. The journey is thus a double structure: one track is the life of bodily engagement with the world; the other track is the life of reflection in order to re-present textually, through images, through signs of all kinds, the experience of the journey.

This textual "re-presentation" of life is what the researcher does and he or she does it within certain limits. I can only re-present life out of the values, linguistic and explanatory frameworks (discourses) available to me. My account of my trip to the dentist will be delineated by these factors. And those who read my account will bring their own baggage to their interpretation. There is no stable meaning to be excavated. Postmodern researchers argue, then, that since all research is a re-presented text which has been carved out of a local context and from the language, values and discourses available to those involved in the research, they should avoid grand, generalizing claims for their work. Instead the research text "becomes an invitation for the creative play of others" Schostack (2002: 230). This perspective creates quite a distance from the scientific tradition.

Law (2004) writes of the "method talk" which he sees to characterize much scientific work. The rhetorical nature of this talk is concealed beneath an underlying claim that the research is value free. We have already seen how this works with the use of the passive to create a scientific air. Additionally, terms such as "robust," "replicable," "validity," "generalization" and "evidence-based" are felt to script science as a stable, singular truth-seeking endeavor. In reality, writes Law (2004: 11) science suffers from an "inheritance of hygiene" problem, namely the denial of the messiness of the human world and of our inevitably partial efforts to understand it. Uncomfortable with this messiness, researchers work up "smooth narratives" (Law 2004) that offer closed truths. This is an overdrawn picture of the scientific tradition, although it is true that there is little problematizing of language within it. Perhaps we need a middle position between radical adherents of the linguistic turn and those who neglect the influence of language entirely.

While it is the case that our research texts can never be more than re-presentations of reality, let us not give language too much determining power. Firstly, some re-presentations are better than others. Silverman's (1997: 239) point that research is more aesthetic than methodological is an important one here:

> Many volumes have been written about how theoretical or political (either implicit or explicit) shape the research task. However, at root, I feel that such positions have relatively little to do with many researchers' sense of what constitutes a "worthwhile" research problem, "interesting data" or "compelling" analysis ... I believe that the most important impulse has more to do with our tacit sense of the sort of appearance or shape of a worthwhile piece of research. In that sense, research is informed by an aesthetic.

Secondly, let us remember that language is facilitative as well as determining. As Vygotsky (1962) pointed out, every word is a microcosm of consciousness. Some adherents to the linguistic turn have become so concerned with the ways in which language might confine explanation, they appear to have neglected the developmental and elastic capacity of language to support our articulation of what might be happening. Language inhabits a space between constraint and possibility.

Thirdly, some reality is knowable in fairly straightforward ways, allowing for observation and language to work together unambiguously. For instance, most researchers are happy to establish that it is raining by seeing or experiencing the wetness of the rain; they are equally happy to describe this wet stuff falling from the sky with the relatively innocent signifier "rain." However, if someone says that it is raining "cats and dogs" or that it is raining in their heart, then we are in a different interpretive league. While social constructionism alerts us to important interpretive issues, we need to appreciate that interpretive challenges are also to do with making judgments about the complexity of the situation.

We can accept social constructionist concerns about the ways in which we make data as much as we gather it together with the concerns for re-presentation discussed. However, I do not think we have to offer endlessly reflexive, negotiated, context sensitive texts in the light of these issues. The point of research is to enable us to make informed judgments about what *might* be going on within and beyond the situations we are researching.

In making these judgments, concerns for complexity and power are also important. Before closing my discussion, I will briefly signal approaches that foreground these concerns.

Critical Theory

"Social research" writes Singh (2007: 12) "is rarely a benign activity, it serves to explain and explain away, and therefore it can serve different functions." Critical theory is commonly associated with research which is sensitive to questions of power of this sort; it also aspires to put research in the service of social justice—here is Cohen et al.'s (2007: 26) definition:

> Its intention is not merely to give an account of society and behavior but to realize a society that is based on equality and democracy for all its members. Its purpose is not merely to understand situations and phenomena but to change them. In particular it seeks to emancipate the disempowered, to redress inequality and to promote individual freedoms within a democratic society.

Similarly, research perspectives from feminism, anti-racist, queer studies and critical realism have a concern for power and social justice. Chapter 9 on Action Research discusses emancipatory models of action research that identify with these approaches. Other chapters pick up on the question of researching marginality where the challenge for researchers is to be sensitive to the experiences of disadvantage while resisting an overdetermined view of those who bear its brunt. Research into the experiences of specific groups of learners is often driven by the search for common characteristics among such groups at the cost of attention to the internal heterogeneity that characterizes any social group. There is, for instance, an idea in circulation among some researchers of international issues that one can speak of "the Chinese" or the "Far East" learner.

Complexity

Bearing in mind Oakley's caution that paradigms are always normative, entering the research setting with an "ism" risks taking an answer rather than a question into the inquiry. Generally, the researcher needs to replace a determination to find a set of common experiences or a single causal link for an openness to a more complex picture. Haggis' (2004) use of complexity theory provides a fresh angle on how we need to be alert to our temptation to search for deep structures and regularities in our data. She advises the researcher to explore local interactions, interconnectedness, singularities and differences to resist this temptation. I have tried to integrate this advice throughout the chapters.

Conclusion

The chapters in this book are informed by a broad interpretivist approach that takes account of the issues of language, power and complexity briefly signalled above. In balancing the theoretical and the practical in the chapters that follow, I hope I have given the non-social scientific reader a sufficient basis for the intelligent application of the methods discussed.

Further Reading

Bentz, V.M. and Shapiro, J. (1998) *Mindful inquiry in social research*. Thousand Oaks, CA: Sage.

Denzin, N.K. and Lincoln, Y.S. (Eds.) (2000) *Handbook of qualitative research*. Thousand Oaks, CA: Sage.

Law, J. (2004) *After method: Mess in social science research*. Abingdon, UK: Routledge.

Seale, C. (1999) *The quality of qualitative research*. London: Sage.

Shank, G.D. (2002) *Qualitative research: A personal skills approach*. Columbus, OH: Merrill Prentice Hall.

Silverman, D. (Ed.) (1997) *Qualitative research: Theory, method and practice*. London: Sage.

2

GENERATING AN
ETHICAL FRAMEWORK

Each research approach I present will have distinctive, as well as overlapping ethical issues, which I discuss in the individual chapters. There are at least two good reasons for having a strong ethical framework for a research project. Firstly, it has a protective function both for the researcher and for the researched. To illustrate, in the summer of 2007, the Australian University Queensland University of Technology found itself at the center of a row concerning the ethical issues related to a PhD thesis entitled: *Laughing at the Disabled: Creating comedy that Confronts, Offends and Entertains*. Whatever the truth, rights and wrongs of the incident, the huge explosion it created (including the suspension of two academics) provides a salutary reminder that we need to think very carefully about the language, messages, tone, intentions, integrity, assumptions and effect on others that our research activity and presentation constructs.

The second good reason for a strong ethical framework is that it is facilitative. An ethical orientation supports the thoughtful conduct of the research process and the eventual credibility of the report. As Shank (2002: 97) simply puts it "a good researcher is an ethical researcher." Such a researcher knows that methods and values are entwined and that any inquiry requires demonstrable respect for the humans within its range. Shank (2002: 97–99) reduces the matter to four essential notions, namely to "do no harm," to "be open," to "be honest" and to "be careful." These capture the spirit of an ethical orientation very well, particularly when wedded to another helpful comment from Shank (2002: 97): "becoming an ethical researcher is a lifelong process. That is, we can never say that we have no more to learn or understand about the ethical implications of our actions."

The ethical dimension, then, should never be viewed as a tiresome preliminary to the real business of research. An early consideration of the ethical framework is also important for the selection of research methods. In particular, ethical issues associated with access to research subjects will be more pronounced for some methods than for others. Further, ethical dimensions will relate specifically to the research question and focus in hand. Clearly, an exploration of bias in marking is going to throw up very different ethical issues to that of an inquiry into library catalog usage. The ethical dimension to research is always situated though there is no shortage of published general guidance to support the formulation of a context sensitive, ethical framework for a particular research project.

General Guidelines

All countries and subject communities have Research Associations which publish ethical guidelines. In the US, there is the American Education Research Association (AERA) and in the UK there is the British Education Research Association (BERA). The following draws on some of the AERA headings to address the particular ethical issues associated with the approaches reflected in this book.

1. Trustworthiness

The AERA (2008: 1) preamble urges that researchers "warrant their research conclusions adequately in a way consistent with the standards of their own theoretical and methodological perspectives." I deal with questions of "trustworthiness" in each of the chapters in this book. One of the important moves for generating trustworthy accounts is to embed degrees of researcher reflexivity into the research. Broadly this means paying attention to where you are coming from (researcher positionality) and how this might influence the conduct and reporting of your research. In the spirit of researcher reflexivity, it will support the generation of an ethical framework if you begin by formulating a statement about the values, experiences, knowledge, interests, beliefs and ambitions which might shape your research interest and focus. For instance, a researcher investigating students' and academics' perceptions of climate change might write something like the following:

> My own position on climate change is that this is a serious problem facing the world and I am committed to campaigns of awareness and political change to reduce the factors which contribute to climate change. While I do not expect to withhold my own feelings about this issue when talking to students or academics, I will try and be as invitational and open as possible to facilitate uninhibited debate about this question.

A further paragraph could deal with what Doucet and Mauthner (2002: 134) have called "knowing responsibility"; to achieve this, they argue:

> that a wide and robust concept of reflexivity should include reflecting on and being accountable about personal, interpersonal, institutional, pragmatic, emotional, theoretical, epistemological and ontological influences on our research, and specifically about our data analysis processes.

Some of this is addressed in the first statement but the following would expand on the epistemological and ontological influences:

> As an environmental scientist I expect research findings to be provisional and contested. Within this understanding, I expect to make judgments on the basis of the weight of the argument; this means taking seriously the evidence for both arguments and counter-arguments; it also means paying attention to who has sponsored the research, the track record of the researchers and any political interests being served by the research. In gathering views of students and academics, I acknowledge that I am bound to influence the data yielded through the way I frame the questions, the method by which I interpret the responses and the research tools I use. I will strive to be reflexive about these issues in order to acknowledge my own place in the research (for this reason I have chosen to write in the first person rather than the passive form that is normally associated with scientific research). Although I am not expecting this research to yield objective truths, I will use strategies such as sending interview transcripts to interviewees for their checking and further comment to maximise the trustworthiness of my evidence.

These kind of initial statements can be treated as provisional and subject to changes or extended at any point throughout the research cycle. Like Brew (2001), I think that researching is a synonym for learning and it would be disappointing if a researcher did not want to add or expand starting-position statements.

Another strategy you can adopt to warrant research conclusions "in a way consistent with the standards" of your epistemological and ontological position is to follow the example of a doctoral student I worked with who presented her own ethical and value position at the beginning of her doctoral thesis. In this way, she explicitly set the standards by which she wanted her work to be judged. Below are some issues you might want to consider in adopting a similar move:

- *Reflexivity*   Can you claim that you have acknowledged your positionality in the research?
- *Curiosity-led*   Have you read your data for the ambiguities, disconfirmations and rival explanations it offers? Have you avoided the forcing of an argument or an answer-driven orientation?
- *Intellectually informed*   Have you developed a strong and convincing engagement with a literature to underpin the theoretical claims made?
- *Data quality*   Can you provide a defence for the adequacy and quality of the data collected?
- *Data presentation*   Have you displayed enough data in the report to support the plausibility of the analysis?
- *Corroboration*   Have you shared your analysis with research subjects and/or other researchers?
- *Research report*   Is the report both engaging and plausible? Can it transport the reader to the research setting?
- *Social responsibility*   How have you addressed issues of equity, quality, academic freedom and respect for other scholars? Does your report protect the rights, dignity and confidentiality of the research participants?

Although you would want to finalise standards that address these issues towards the end of the research cycle, an initial set might act as a helpful reminder of the ways in which the ethical and method-ological link together.

If your research is within a postmodern frame, you will want to emphasize the situated nature of moral issues influencing research rather than be guided by overarching general principles (such principles are denounced as grand narratives by postmodernists). You will also want to pay particular attention to overt and covert issues of power, marginality, absences, researcher authority and research audiences/readers. See Grbich (2004: 88–93) for a succinct and helpful discussion of these questions.

2. Research Participants and Participating Institutions

Again to quote from the AERA (2008: 1), this principle concerns respect for individuals and institutions:

a) *the rights, privacy, dignity and sensitivities of their (researchers) research populations*;

b) *the integrity of the institutions within which the research occurs.*

Understandably, the AERA is particularly concerned about the ethical issues relating to researching children in schools and other institutions. While the power issues between university teachers and students are less dramatic than those between pupils and researchers, it is important for researchers to be mindful of their gatekeeping power over students and ways in which this will influence learner contributions and dis-closure. We cannot design out power from our research but we can acknowledge its presence and do our best to minimize its effect. Clearly researching *with* students is one way of doing this.

In terms of respect for the integrity of institutions, in my experience this is often overlooked because our focus tends to be on the individuals within them. When departments and institutions open their doors to researchers, they trust that their activities will respect their right to representative, constructive and confidential (where necessary) findings which are protective of both individuals and the institution as a whole. This is particularly important for evaluation research, which I discuss below.

I have been involved in the commissioning of research where the institution was unhappy with the findings and quite rightly asked to see the substantive evidence in relation to particular claims. Since the evidence was rather patchy, the institution insisted that the entire report be withdrawn and deleted—this option had been laid down in the ethical agreement established with the institution. It is important to be clear that in this case the university did not object to public debate about sensitive issues but it did not want this debate to rest on unsatisfactory evidence. The cautionary tale here is to pay careful attention to the quality of data, particularly where it relates to sensitive issues. Do not be so driven to tell a particular story that you end up shoehorning your data into it. This is narrative fraud.

My cautionary tale also alerts us to the importance of the ethical at the data analysis stage. Glucksmann (1994: 163) reminds us of this importance in relation to the analysis of interview transcripts:

> Usually the researcher has sole access to and total control over the tape or transcripts. No one else overseas which parts she selects as of significance . . . each researcher is left on trust to draw the difficult line between interpreting the data in terms of its relevance to her research questions as opposed to twisting it in a way that amounted to a mis-representation of what was said.
>
> (Glucksmann, 1994: 163 in Maynard and Purvis: 129)

Sometimes researchers return transcripts for the checking of their accuracy whereas Glucksmann's point suggests that it is much more meaningful to send the provisional analysis to the participants. As Walford (2001) has pointed out, nailing participants to what they "actually said" when you interviewed them is not proof that an authentic voice has been captured. They may have responded on the hoof and want to reject or revise what they said. What matters is what you have made of what they said and what they think of this making. The transcript is not an affidavit.

3. Informed Consent

This is about ensuring that participants in your research are aware of its purposes and their role within it. Informed consent at the beginning of a research process might need renegotiating as the research proceeds so that the participant is made aware of the emerging exploration and analysis. As Duncombe and Jessop (2002: 111) write in relation to the interview participant "it is clearly impossible to give their fully informed consent at the outset of an essentially exploratory qualitative interview whose direction and potential revelations cannot be anticipated."

Denscombe (2007: 146, 147) provides a helpful guide to the construction of a consent form; this is signed and dated by the researcher and the participants at the beginning of any research project. Here are some of the key elements it addresses:

- Brief details of the research project (aims, methods, anticipated outcomes and benefits).
- Broad ethical code (usually associated with Subject Associations such as AERA).
- Contact details.
- Expected participant contribution (and any rewards for participation).
- The right to withdraw consent.
- Confidentiality and security of the data.

The last point also covers data protection legislation issues. Denscombe (2007) points out that the best way to avoid getting involved in legislative frameworks is to ensure that the data is anonymous and thus untraceable to whoever provided it; this ensures that the data remains outside of the legislation. Naturally, the data has to be genuinely anonymous and this becomes tricky in cases where the research population is small or where particular participants can be detected. For instance, interviews with female or ethnic minority Vice Chancellors in the UK would be based on a rather small segment of the UK Vice Chancellor population. In this event, guarantees of anonymity cannot be made confidently and some other means will need to be explored to protect the privacy of participants.

One of the difficulties with standard consent forms is that they are designed for research that is largely "extractive" of information from participants. Given the spirit of the research approaches in this book to acknowledge the negotiated character of many qualitative

research encounters, perhaps under "expected participant contribution," some of the detail might read:

> Participants will be asked to explore with the researcher their experiences
> of undergraduate assessment. Participants will be invited to contact the
> researcher at any time if they have any further comments or thoughts
> to add. A provisional analysis of the research will be sent to the
> participants so that they have a further opportunity to contribute to the
> exploration.

Most researchers are aware that "informed consent" is a problematic
notion, especially where there is uneven knowledge and power between
researcher and respondent. This is another reason why a notion of
ongoing, negotiated consent might be appropriate. This usually
involves sharing provisional findings as part of a process of negotiating
meanings with participants.

There are a few cases in which deception is necessary to the
research design and this has to be strenuously defended before Ethics
Committees will clear it. They will want to know, for instance, the
likely reaction of participants once the deception is made clear to
them. If the revelation of the real purpose of the research is likely
to cause discomfort, a Committee will be far less likely to grant
clearance.

Ethnographic approaches involve observations of some kind. Covert
observation is particularly problematic because it is a form of deception.
It is worth remembering that a novelist's field notes are full of character
sketches based on observations of real people and real settings. But
novelists then pick and mix from these sketches, producing characters
that cannot be traced to real people. Indeed, authors often declare
that their characters are purely fictitious (a half truth). Since this
pick and mix ploy is not available to the researcher (perhaps it should
be?), covert observation is generally frowned upon unless a very good
defense can be mounted for it. That said, there is a level at which
any kind of observation has covert elements because people often
forget that they are being watched even when they have consented
to it. Moreover, how can they know in advance what you will pay
attention to and whether they are going to be comfortable with your
focus? These questions are situated and best addressed accordingly.

Internet research has prompted fresh concern about privacy and deception. The opportunities for researchers to "lurk" in chat rooms has created new questions. Should researchers announce themselves to the chatting groups? Should the fact that the discussions are already in the public domain license the research? Given that people chatting often have anonymized themselves through "handles" that are not traceable, how can the research threaten their privacy? Again, it seems that these questions are best answered in situated ways. For instance, Denscombe (2007: 149) offers the example of researchers getting clearance for exploring white racist chat rooms; partly this clearance was to do with the public availability of the data and partly through an assurance that participants' identities would be protected.

4. Do No Harm

Connected to the last point in particular but also as a general principle, the research imperative "to do no harm" is a central one for ethical approval. In education, this relates to the need to ensure that a research intervention does not disadvantage a group, particularly at the point of attainment. Similarly, the intervention must not unduly privilege a group although sensible decisions have to be made about this. If, for instance, five academics teach the same module and one academic decides she or he wants to trial a more interactive pedagogic style, this is also within the bounds of professional discretion. And whether she or he turns this into a research project or not is irrelevant. As we know, variation in module marks can sometimes be tracked to gifted, committed teachers at one end and less gifted ones at the other end. If, on the other hand, the department decides to restrict extra tuition to just one cohort of students studying this module, this would be ethically difficult to defend (this is not, of course, the same as offering a student support service which provides extra tuition on demand).

5. Evaluation Research

Evaluation research does not always need clearance if it sits within the spirit of an institutional commitment to securing student feedback and continual improvement. Thus an additional mid course evaluation to supplement institution or faculty-wide end of course evaluations is unlikely to need ethical clearance so long as it is for development purposes. If, on the other hand, the data is to be used in public research, then clearance needs to be sought.

Commissioned evaluators have particular challenges because they work for diverse stakeholders, including fundholders, who may have conflicting agendas and experiences of an initiative or program. In the extract below, Stake is discussing qualitative research generally but I think his comments have strong application to evaluation research:

> privacy is always at risk. Entrapment is regularly on the horizon as the researcher, although a dedicated non-interventionist, raises questions and options previously not considered by the respondent . . . some of us "go native" accommodating to the viewpoint and evaluation of the people at the site—then revert, reacting less in their favor when back again with academic colleagues.

> (Stake, 1995: 46)

These tensions signalled by Stake certainly resonate with my experience. Sometimes there is pressure from the commissioners to portray an initiative in a good light or to "find" in a particular direction to support a desired policy shift. Some people involved in a program or initiative have a personal agenda to progress or have complaints about management to air. The chapter on Evaluation focusses on evaluation stances to support thinking about these questions. Another problem associated with external evaluations concerns the quality of the design and conduct. I have seen evaluations where data is snatched, grabbed and fashioned into a story in unseemly haste; this is often because funders have limited resources to support a thorough going evaluation or the work has gone to someone who has insufficient time available for the work. A quick and dirty evaluation is worse than no evaluation.

In recognizing the ethical complexity of evaluation methods and settings, the UK Evaluation Society offers very extensive (and a little overwhelming) guidelines. To support the generation of an ethical framework for the evaluation of teaching and learning programs, I participated in an expert panel convened by the Higher Education Funding Council for England—many of the points below are drawn from my interpretation of some of the panel's discussion. Most of the headings are extracted from the UK Evaluation Society's guidelines (n.d.).

1. *Demonstrate that the evaluation design and conduct are transparent and fit for purpose.*

   One way in which this demonstration can be made is by getting other professionals to review the evaluation framework (most evaluation reports offer an appendix which sets out the design and method of the research).

2. *Where possible, respect both the audit requirements and development dimensions to programmes and initiatives. Work with commissioners and stakeholders to agree how the evaluation can integrate these summative and formative aspects.*

   There is often a need for evaluations to have the dual purpose of coming to some judgment about the worth of a program or initiative and of supporting its development. Some evaluators privilege one at the expense of the other and this principle is an appeal for them to be mindful of both of these purposes. In my experience, even where the thrust of an interim evaluation is formative, stakeholders do want some way of measuring their influence for an audit sensitive audience.

3. *Attend to the need for stakeholders to achieve, where possible, an economy of effort between their monitoring and evaluation activities.*

   Evaluators often place undue burdens on stakeholders to provide evaluative data. This is a plea for them to rely, where possible, on monitoring and evaluative data that is already collected by stakeholders where this seems appropriate. Indeed, a first question for an evaluator needs to be "what naturally arising data can I use?" While dedicated data is likely to be needed, a consideration of what is already in the system is an important initial one.

4. *Demonstrate a commitment to the integrity of the process of evaluation and its purpose to increase learning in the public domain.*

This is about sustaining a developmental thrust to evaluation so that all those involved learn from the process more than they are judged. This requires inclusive moves such as sharing analysis, talking to everyone (within reason) and avoiding undue judgmentalism. This also requires that the evaluator writes and communicates findings in accessible and interesting language.

5. *Avoid client pleasing by adopting a reflexive distance from stakeholder interests. Do not evaluate with one eye on the next contract.*

   This is a reminder that evaluation research is often a commercial business and the imperatives of competition can compromise the integrity of the work.

6. *Be aware of and make every attempt to minimize any potential harmful effects of the evaluation prejudicing the status, position or careers of participants. Ensure there is follow-up to deal with sensitive issues which may be raised as a result of the evaluation.*

   The judgments made by evaluators can become the last word, influencing the direction of careers, organizations and learners. Evaluators need to be sensitive to this potential influence and where relevant ensure that their conclusions are judicious, helpful and evidence supported, as the next principle outlines.

7. *Demonstrate comprehensive and appropriate use of all the evidence so that conclusions, recommendations and discussion can be traced to this evidence.*

   This is self-explanatory and clearly links to the preceding points.

8. *Clearly associate the evaluation research with guidelines such as those of the British Education Research Association.*

   This simply requires the evaluation design to declare its ethical framework and to associate this with general principles such as those set out by BERA or AERA.

9. *Work within Data Protection legislation and have procedures which ensure the secure storage of data.*

10. *Acknowledge intellectual property and the work of others.*

Conclusion

I will conclude with Miller and Bell's (2002: 53) insistence that:

> satisfactorily completing an ethics form at the beginning of a study
> and/or obtaining ethics approval does not mean that ethical issues can
> be forgotten, rather ethical considerations should form an ongoing part
> of the research.

I will touch on ethical issues throughout the chapters and hopefully
this chapter serves as a framework for them.

Further Reading

American Educational Research Association (n.d.) Ethical standards. Available
online at: www.aera.net/uploadedFiles/About_AERA/Ethical_Standards/
EthicalStandards.pdf.
British Educational Research Association (BERA) (n.d.) Available online at:
www.bera.ac.uk/.
Denscome, M. (2007) *The good research guide*. Maidenhead, UK: Open
University Press.
Mauthner, M., Birch, M., Jessop, J., and Miller, T. (2002) *Ethics in qualitative
research*. London: Sage.
UK Evaluation Society (n.d.) Available online at: www.evaluation.org.uk/.

3

QUALITATIVE DATA ANALYSIS

The Appeal

Qualitative analysis enables the researcher to: a) get at complex layers of meaning from research texts or visual data; b) interpret human behavior and experiences beyond their surface appearances; c) provide vivid, illuminative and substantive evidence of such behavior and experiences; d) build theory inductively from qualitative data sources.

Purpose

Qualitative data analysis explores themes, patterns, stories, narrative structure and language within research texts (interview transcripts, field notes, documents, visual data, etc.) in order to interpret meanings and to generate rich depictions of research settings.

In much qualitative research, data gathering and analysis are dynamically linked. The purposes of entwining the two are to enable manageability of the data, to allow for continual focussing of the inquiry and to generate theoretical insights. Further, decisions about data analysis need to be made at the beginning of a research project because these involve decisions about researcher positionality and the knowledge claims you want to make for your research.

Finally, the purpose of data analysis is to mutually engage empirical data with a theoretical literature and with the researcher's reflections (usually recorded in a research diary). Thus, it is important to bear in mind that the material for qualitative data analysis is not simply the "raw" gathered data.

Theoretical Concerns

Data as Driftwood?

Data, writes Schostack (2006: 68), is not "something like a found object on the beach, a piece of driftwood." Many contemporary qualitative analysts, like Schostack, now accept that however they have gathered it, their data can never be neutral. Data gathering is always a selective process in which we privilege some sources and discard or exclude others. Most qualitative researchers also accept that their analysis and write-up are deeply influenced by their own positionality.

To say that data is never objective is not to suggest that qualitative analysis produces hopelessly arbitrary accounts. While there is no formulaic prescription for the production of trustworthy research reports, generally, these involve a reflexive and thoughtful engagement with the data gathered and of the literature read. Above all, there needs to be an appreciation of Glaser and Strauss' (1967: 251) proposal that "the root source of all significant theorizing is the sensitive insights of the observer himself."

Technique and Art

Qualitative data analysis involves techniques but it is not a technicist operation. In their excellent book on qualitative data analysis, Coffey and Atkinson (1996: 10) write that data analysis is "imaginative, artful, flexible and reflexive" as well as "methodical, scholarly and intellectually rigorous." Bringing these qualities to bear on data analysis involves forms of serious play (Stronach and McClure, 1997) to include a respect for but not slavish adherence to the perspectives on offer. In this spirit, we might also replace the terms "techniques," "procedures" and "strategy" for that of "moves" to capture a more cultural and creative sense of what we are doing when we analyze data.

Data Reduction

In one way or another, all data analysis perspectives offer moves that achieve some form of data reduction: at one end these moves codify and segment research texts and at the other end, there is an attempt to pull out emergent stories from the text within a holistic approach. These opposite ends express a tension between the need to reduce the research text for intelligibility and the need to maintain its integrity so as not to do violence to people's testimonies. Managing this tension involves swinging from chunking the data to looking at it as a whole.

Grounded Theory

Early theorists of qualitative analysis come from the social interactionist tradition clustered around Chicago University. In particular, we are indebted to Glaser and Strauss' (1967) elaboration of how to analyze qualitative data in their seminal book, *The Discovery of Grounded Theory*. Glaser and Strauss later parted company because they held different views about how to progress their theory and procedures. See Grbich (2007: 74/83) for a clear discussion of their subsequent differences and of grounded theory generally. I remain enthusiastic about Glaser and Strauss' original text as a source of inspiration for qualitative data analysis and below I will draw on some moves suggested by these authors for coding textual data though I do not claim to offer a faithful version of grounded theory.

An Inductive Approach

Those who are unfamiliar with qualitative data analysis often find it hard to see how a mass of prose and/or visual data can be reliable evidence. In their first try, many are tempted to limit their organization of the data to the descriptive, clustering it around the questions which have been asked or according to a hypothesis being pursued.

Following Glaser and Strauss (1967), at least two moves can take place to avoid a descriptive approach. Firstly, the researcher needs to set aside what she/he is looking for and try and work out what the data is "telling him or her." While this will always be an interpretive act, it is an inductive stance that will encourage researchers to focus on what the researched might be saying rather than on their question and interest. Even if you gather data on the basis of a hypothesis, you do not allow this to determine what you look for; of course it is likely to hover in the back of your mind as you explore the data but the spirit of the inductive approach is to be researched-centered rather than researcher-centered. You are trying to look at what *they* say rather than what you want them to say. You are exploring a theoretical direction *from* the data rather than theory testing in which you risk forcing the data to tell you what you are looking for.

Secondly, as I next discuss, whatever method of analysis they use, researchers need to have a way of thinking about what the data may be saying in relation to their *reading* of a literature. The importance of connecting analysis with a scholarly engagement of the literature cannot be stressed enough.

Empirical and Theoretical

There are three key sources for qualitative data analysis, namely the gathered empirical data (e.g. documents, interview texts), the researcher's reflective notes and a literature. The literature can be a combination of explicitly relevant academic texts as well as novels, films, documentaries, news cuttings, etc.—anything that supports the advancement of understanding of the issue in hand. Avoid a linear process in which a literature review precedes analysis of the empirical data. They are best seen as feeding into each other throughout the research cycle: read a literature as you analyze and analyze as you read a literature. This dynamic allows you to approach the analysis of data (and indeed its collection) with sensitizing concepts.

Sensitizing Concepts

This is another Grounded Theory concept, deriving, originally from Blumer (1954). Sensitizing concepts support your provisional research question by helping you to steer the direction of your inquiry and your theorizing. Say, you have started to observe a seminar series in pursuit of a provisional research question: "*how is learning made possible by the seminar method?*" As you conduct your research, you come across a literature which demonstrates that a teacher's expectations of a learner will influence how the learner performs. This dynamic is known as the self-fulfilling prophecy. Let us suppose that this concept resonates with some of your observations where you note that the teacher asks questions of the same group of students. In this event, "the self-fulfilling prophecy" becomes a sensitizing concept, allowing you to focus your inquiry. You are now curious about whether the group of students receiving the most attention are also getting the best grades. Your further inquiries might yield a mixed picture, suggesting that the "self-fulfilling prophecy" does not entirely explain the picture. Perhaps you adapt or discard the concept. Certainly, you will want to carry on reading as you collect and analyze the data to keep an eye out for further promising theoretical leads.

Trustworthiness

Trying to make qualitative data analysis reliable in ways that wholly emulate quantitative data analysis is like trying to make apples taste like oranges. There is a numerical dimension to qualitative analysis but the important point to stress here is that qualitative researchers strive to shed light on questions that simply cannot be answered by surface observation or by statistical analysis alone. The qualitative research enterprise is to get at complex layers of human meaning through interpretive moves described below.

Issues of trustworthiness are handled by most analysts of qualitative data by taking a reflexive stance as described next. This stance accepts that the subjectivity of the researcher will always be present (this equally applies to quantitative research) and the best way of addressing this is to openly engage with it rather than to assume the unreachable posture of objectivity.

Methods

Reflexivity

Firstly, here are a set of suggested principles to support a reflexive approach to data analysis:

1. Analyze as you gather data—do not leave it to the end.
2. Approach the data inductively—what does it appear to be telling you?

3. Address a tension between the inductive ideal (in 2 above) and the influence of your interpretive baggage.
4. Find ways of coding data that avoids over segmentation.
5. Memo as you code to capture reflections and theoretical possibilities.
6. Read other research/theories as you look at the data.
7. Avoid premature closure or the enforced categorizing of data.
8. Explore disconfirming as well as confirming evidence.
9. Avoid cherry picking the quotes that confirm what you want to say.
10. Explore patterns of frequency, recurrence and absences.
11. Explore what the singular and aberrant might tell you.
12. Think about what is being said alongside how it is being said and who is saying it.
13. Think about language—figures of speech, structure, idioms.
14. Think about what is not said—the silences and absences.
15. Think about your own positionality in the research.

This is not an exhaustive list, of course, but it aims to capture in summary form the analytical moves I now discuss.

Gather and Analyze

As you gather, you provisionally analyze to achieve manageability of the data. Straddling the two activities will also support the determination of promising leads both for the gathering of further data and for the identification of supportive reading and conceptual development.

Get Intimate with the Data

It is important that you read your data a number of times over to review first impressions; if you are dealing with transcripts, in your first stage of analysis you can type them up yourself. If you get someone else to do the typing, listen to your tapes as you read the transcripts. Ensure you include paralinguistic details (sighs, silences, laughter, etc.) and keep to the language of the respondents. Whether it is transcripts, documents, videos, field notes or other forms, the frequent contemplation of your data needs to take place throughout the research cycle. It is also wise to discuss your emerging analysis with whoever is willing to listen and contribute to it.

The Fetish of Transcription?

In his helpful and very readable book, *Doing Qualitative Educational Research* Walford (2001: 92) argues that we have come to fetishize the interview transcript. We represent our detailed transcriptions as a stable, accurate record without regard to the possibility that what people say on a hot Monday afternoon might be different to what they say on a Friday morning. In this way an interviewee comment, writes Walford (2001: 97) might be "wrenched out of its context and presented as if it represented the 'truth' about one person's views or understandings." This is always going to be an interpretive problem even if you do not transcribe the interview, but Walford has a point. He suggests that you listen to taped interviews several times and take notes, capturing verbatim, selected elements for the report (a similar move can be adopted for the analysis of video). In defence of the whole transcript approach, in my experience, the sheer bulk of a set of transcripts invites an analytical, meditative, intimate engagement with them. In this sense, note taking from tapes or videos changes the nature of the intellectual labor involved in data analysis. You might want to try out both methods to explore their differences.

Coding

Coding will allow you to start thinking systematically about what the data might be telling you. It will also begin the process of theory generation.

I will illustrate how coding might proceed from the following interview extracts from UK students talking about going to their family home during their first year at university (Fine, 2007). If you are doing this manually, you will need several copies of each transcript because you are likely to cut up the same segments for different themes. If you use software, it will allow you to code the same segment as many times as you want to. This is a key advantage of using software—it does save you having lots of bits of paper arranged across your dining room floor.

Whether you use software or manual analysis, you will need to tag your research texts carefully so that you do not separate segments from their source. You will want to make decisions about how to tag and what level of source information to keep. For interview transcripts, decisions will concern using pseudonyms, "m" for male, "sss" for social science student, etc. For field and diary notes, you will want to ensure that tagged segments include contextual information such as date and place they were written.

In the first instance, you want to simply look at what the respondent is saying and formulate ways of chunking it. This is a fairly descriptive phase. Where feasible, it is good to use the language of the respondents for your initial coding because it helps you to stay close to the text. You are basically looking at each comment or set of comments and asking yourself "what is this about?" (Dick, n.d.). You then decide on a provisional label to capture this and assign relevant comments to it. Here are four interview transcript extracts to illustrate:

1. *Erm being away from home was*
   kind of different because you don't
   have that sort of family security,   Family/security
   you replace it really quickly with your
   new friends, so that wasn't too bad   New friends/
   at all. Cos you keep yourself so busy   security
   at uni, you don't really stop and think
   oh I miss my parents. Returning home,   Missing parents
   erm it was really nice, apart from the
   fact that you have to adjust back to the
   model of the family, the way they kind
   of have, not the final say, but you have   Adapting to
   to kind of give in to their views. Rather   home
   than put yourself first, you have to be a
   bit more like, oh ok dad, just to please
   them, to keep the family kind of functional.

2. *Yeah well we went for two and a half
weeks and then I had like three days*   Going home
*in the family home when I got back
before I came back to Hampton. That
was crazy, it was really emotionally*   Emotionally
unsettling. I had like three days to see   unsettling
all my friends and relax back into family   Push/pull
*life, while having to pack up everything
for Hampton and it was just a funny
place to be in your head. I found it
really stressful. I just wanted to relax*   Relaxing at home
with my cats and nice food.

3. *Yeah it was good to get my mum to buy
shopping for me, definitely lots of groceries* Going home
*erm and it was good to, like there are
certain things you miss like having a
massive bed. I mean it's sort of like*   Home as holiday
*being on holiday, not having everything
with you. I took the train back so I didn't
have all my clothes. So yeah it was
definitely a lot more like a holiday than
I expected it to be. Not sort of a going
away holiday but one of those boring*   Home boring
*ones you go on with your family, in
a caravan or something, which is nice,
cos you get away, but it's also really dull.*

4. *I like going home, but I like coming back*   Going home
*to university as well. I felt like it should
be oh I love university and I don't wanna go
back or the other way around but it's not. I like
both. I don't mind either prospect. They're
both just as good as each other for different* Uni/home
reasons. My closest friends seem to think   balance
*the same, they love, cherish being away
at university as much as to be at home,
they like both too.*

Don't worry about the rough and ready nature of the analysis at this stage. Follow Delamont's (2002: 17) advice to "be wild" by coding data densely and speculatively in your first readings. As you read the data, you will be building new codes and revising former ones. You do not have to code every word or sentence but if bits of the text look like they might be worth saving, put them in a "miscellaneous" code and then look at them again when you have finished the whole text to see if you want to assign them. Some chunks of text will be assigned to a number of different codes. Indeed if you are not sure about where a comment might go between different codes, put them in all of them. If you have taken Walford's advice and done some of the preliminary analysis by note-taking from the tape, your typed up version of the notes will form your research text. Arguably, software is less useful for this method of analysis.

Core and Sub-Categories

Following Glaser and Strauss (1967), your next step is to aim for the identification of "core categories." If, for instance, "family security" is a recurrent explanation about the benefits of going home, this can become a core category because the data appears to be saying that it is a key factor in the experiences you are researching. Generally, you want to end up with a manageable number of core categories that seem to capture what is going on. In selecting core categories you are choosing what seems to be critical to an experience rather than what might be incidental to it.

Sub-categories are properties of the core categories. For instance, you might have a sub-category to "home security" called "dirty laundry" because a few of your respondents said that they took their laundry home to get washed. In the examples above, the experiences of going home are described as "boring" and "relaxing"—these are also sub-categories to which you assign the relevant passages. More analysis of the transcripts from this research led to "adapting to home" as a core category. A lot of the students talked about fitting in with family rules, being treated as more grown up and having more independence—again these are sub-categories to this core one. Further core categories were "home/university" which captured the many comments about the push/pull, unsettling or complementary character of the two "homes." "University friends" provided another important core category. Further analysis of the transcripts revealed that students treated university as a place where they could be more adventurous and experiment with different ideas and friendships. Comments clustered around this theme produced the core category of "experimentation."

Through a process of comparing and re-reading the data for what seems key, also pay attention to the marginal, aberrant, singular and absent because frequency is only one possible measure of importance. Do not rest the credibility of your account on the fact that because many people talked about "x" then "x" must be true. For instance, apparent unanimity of voice about an experience might stem from common experiences and a divergent view might come from someone who does not share this experience (you might have to get new data to check this out). Below I quote from one student who found going home much more difficult than the other students interviewed and this seemed to be tied into his failure to make good friends in his first semester. This single testimony offered a fresh angle on the complementarity of home and university, suggesting the importance of an equilibrium across both homes. Similarly, those students who had boyfriends or girlfriends in other cities found it much harder to settle into university life and this was important to note.

Including Yourself

Conventional analysis of interview transcripts take what is said by respondents. You might want to add a contemporary twist to this convention by including, where relevant, chunks of dialogue under your codes to reflect an awareness of your own contribution to the developmental and negotiated character of the discussion or observation.

Constant Comparison

As you explore the transcripts, you are engaging in a process of constant comparison at various levels. Firstly, you are looking at descriptive categories across texts; then you are looking for what seem to be key experiences (core categories) and what characterizes them (sub-categories). Next, you will want to look at possible linkages across the categories and creating conceptual hooks that explain these linkages. In Fine's (2007) research it seemed that the experiences of going home and of returning to university for first year students are deeply relational and consequential for self-growth. To note these links and related ideas, adopt the grounded theory method of memoing.

Memoing

Memoing allows you to think about links between categories and to connect them to possible sensitizing concepts or to raise questions you have yet to explore. For instance, against a category of "university friends" you might want to note: *are friends at university central because they represent a "family" away from home? What about international students? Are their experiences different?* Later you might want to add to this memo something like "*Turner's work on transition may be relevant*" because you have been reading up on his work on *rite de passage* to help you make sense of what the students are telling you about their transition to university. You may also tag a memo to the sub-category "dirty laundry" that indicates that male students have mentioned this. Theory building is not a mystical process; it emerges from interpretive interaction with the data and the literature in these ways. You are thinking with the data and you can also use memos to register thoughts about your own positionality with respect to the categories and linkages you are generating, e.g. "*is there a risk here that I am overdrawing on my own undergraduate experiences—do I need to ask further questions?*"

If you are working with software, it will enable you to memo against any chunk of data; if you are working with paper, you may want to record categories on the left hand side of your notepaper and allow space on the right hand side to make notes. That said, not all memos will be tagged to particular categories; some will be general observations about where you think your analysis is going or what you might need to do next. You might, for instance, need to return to the field to get more data on an emerging theme.

In thinking about your core categories and their possible connectedness to each other alongside the literature, you will begin to generate conceptual hooks. These hooks allow for abstraction and explanation from the text, as Glaser and Strauss (1967: 23) put it "a concept is a theoretical abstraction about what is going on." To illustrate, in looking at the connectedness of the core categories of "family security" and "university friends," you might generate a conceptual hook of "safe transition" to link their apparent complementary function for student self-growth.

The analytical moves described thus far have generated a promising theoretical explanation of the relational meanings of home and university for many first year undergraduates. In reality, of course, you would be relying on much more evidence and thinking from a larger data set and you would produce more penetrating theory building than I can offer in a short worked example.

Member Checking

If you can, show some of your data to other researchers to see what they read from it in order to add to the plausibility of your account. If you can choose someone who can read the data from an alternative set of experiences, so much the better. I now turn to other ways in which data can be analyzed.

Patterned Data: Content Analysis

If the majority of a cohort of psychology students you have interviewed report a concern that there are too many assessed assignments to do every semester, you can confidently record this as a common concern, not least because you can check this against the course documentation. The data here could be described as patterned because a critical number of students are saying the same thing and the thing is fairly easily identified. Analysts often use content analysis to show up patterns of this sort.

Content analysts are interested in identifying frequency of particular terms, themes, explanations, etc. found in the text. This can be accomplished by "a systematic coding and categorising approach"

Grbich (2007: 112) that attends to frequency and patterns in data. Cohen et al. (2007: 476) reduce the processes of content analysis to the following basic steps:

- breaking down texts into units of analysis;
- undertaking statistical analysis of the units;
- presenting the analysis in as economical a form as possible.

Content analysis, then, involves deciding on what you want to count in a sample of comparable texts (documents, transcripts, etc.)—this might be particular terms or expressions, coded themes or perhaps your own categorization of people's accounts. Cohen et al. (2007: 483, 486, 487) provide a very helpful worked example of how content analysis works. Once you have done your classifying and counting, you present it graphically to give yourself and the reader an at a glance sense of the scale and density of the issue in hand. Then you can move to the analytical stage by thinking about what the identified patterns, frequencies and categorizations might mean. In so thinking, revisit how you derived your categories to check whether you have forced the classifications you made—this is a temptation to resist in content analysis. Make sure you are not neglecting possible disconfirming evidence.

If you do decide to undertake content analysis of your data, a qualitative data analysis software package will be of great assistance. An alternative data move you might want to consider is that of conversation analysis.

Conversation Analysis

Conversation analysis "is a field that focuses heavily on issues of meaning and context in interaction" (Heritage 1997: 162). It can only be undertaken, of course, if you can produce a transcription of a naturally occurring conversation (e.g. telephone calls, television or radio interviews, email exchanges, plays, recorded data). Grbich (2007: 137) explains the scope of conversation analysis:

> In studying conversations every aspect becomes significant from the smallest pause to the loudest yell and all that goes in between . . . what happens when a person of higher status talks to a person of lower

status? How do conversations between women and between men differ from those when the genders are mixed? How do we manage both our own and others' emotions and impressions? For example, death can be discussed on a continuum from the light hearted parrot sketch of "Monty Python"—He is no more, he is an ex-parrot"—to the unctuous "He is in a better place" of the funeral director.

In exploring these kind of issues, Heritage (1997: 164) outlines six areas of attention for the conversation analyst:

- *Turn-taking*   The study of who speaks when can shed light on a range of conventions and power issues in conversation.
- *Overall structural organization of the interaction*   Attention to the structure of a conversation will allow an understanding of the "task orientation" of the exchange.
- *Sequence organization*   Allows us to see "how courses of action are initiated and progressed" (Heritage 1997: 169).
- *Turn design*   This is where a conversation contribution is aimed at determining the subject and the response.
- *Lexical choice*   This concerns the terms we use to signal the tone and purpose of the conversation. Starting a conversation with "hiya" establishes an informality which would contrast with, say, "hello, this is Dr. Smith from the hospital."
- *Interactional asymmetries*   As the term implies, this is where there is a power and/or purpose differential between the parties of a conversation.

Details of this method are succinctly presented in Heritage (1997) and in Grbich (2007).

Narrative Analysis

One way of avoiding over-fragmenting the data is to look at its overall narrative structure. As Coffey and Atkinson (1996: 69) write "How social actors retell their life experiences as stories can provide insight into the characters, events and happenings central to those experiences."

In order to conduct a narrative analysis, the data has to provide a story structure—this crucially involves an account which has a

LIBRARY, UNIVERSITY OF CHESTER

beginning, a middle and an end. See the chapter on Narrative Inquiry but to illustrate the usefulness of this approach here, let us look at the following story-like extract from another student talking about home from Fine's (2007) research:

> I've moved into an area and a house that I really love. So I think home is the most important thing in your life, and I've recently found something that I love living in, the sense of comfort and me being mature enough to say home is important, and the people you live with is so important, now I live with people that I love, but in first year I lived with people that I hated and it made me wanna drop out of uni, I really I tried to move house half way through my first year, I didn't but I tried, if I had it would have been sooo stressful. There were points in my first year when I did find my home town quite depressing when I came home and no-one was there during term time, because everyone was at uni doing their own thing, so it was very desolate. My parent's home was a bit depressing and during my first year when I didn't get on with my housemates, that was a bit depressing. There was definitely a point where both were depressing and there wasn't anywhere where I felt that comfortable. Now I feel like my relationship with my parents has got better, cos they understand me as someone that is doing their own thing, a personality has been formed and moulded even more so, the way I live my life is mine now, my lifestyle choices that I make are mine, and they understand that.

In this story about finding a home from home, we can see how this student's personal growth depended on the right housemates and how finding them allowed him to "form and mould" himself. Looking at this respondent's story, we can conceptualize undergraduate life as a journey, the success of which appears to depend upon successful anchorage away from home. We can also see from this narrative how deeply troubling and precarious this undergraduate journey can be (note the frequent use of "depressing").

Enthusiasts of narrative analysis are drawn to its holistic way of understanding experiences of this kind. When you come to make sense of stories such as this, you will want to think about *how* the narrators have constructed them as much as what they have said. Although this is a crude opposition, it could be said that coding, such as in grounded theory approaches, tends to focus on *what* people say while narrative analysis is also concerned with *how* people say it. Here are some of the "how's" you might want to explore in your data (whether you chunk it or not). Some of these moves come from the field of discourse analysis (see Potter, 1997).

Contrastive Rhetoric

After Hargreaves (1984), Coffey and Anderson (1996: 45) suggest attention can be paid to the contrastive rhetoric people use. Thus, in the quote above, the respondent offers a before/after account in which he counter-poses his move to people he loves from people he hated and from the parental home to home at "uni." His sense-making appears to operate through these binary opposites. Interrogating texts for this form of rhetoric allows us to understand something about how people structure their experiences, views and choices.

Opposition Talk

This is where people contrast themselves to others. In the chapter on narrative inquiry I give the example of a student who reproaches his housemates for their "messiness." Through his oppositional talk (Savin-Baden, 2004), he has positioned himself as the tidy, mature member of the household and distanced himself from the "noxious identity" of messy. Opposition talk can often proceed through the adoption of explicit membership categories.

Membership Categorization

Baker (1997) argues that we often talk about ourselves and others through membership categories. When people use categories about others, they have assigned characteristics to these people. For instance, when I interviewed a teacher of psychology in a prestigious university, at one point he said "we teach to the brightest students." This small statement conveys a lot of meaning. First that the teacher attributes brightness to some students and by implication, non-brightness to those who are not members of this category; second that he positions himself and colleagues as maintaining/defending an exclusive pedagogic standard and cultural practice. When you come to analyze what people say, attention to the categories they use both for themselves and others can be very revealing.

Stake Inoculation

In exploring a television interview with the late Princess Diana, Potter (1997) shows how her use of "I dunno" at certain points are not necessarily innocent expressions of uncertainty. In offering a view as to why her biography caused a stir in the royal palace, Diana begins with an "I dunno" in order to both say what she wants to say (that the palace does not like strong women) and to inoculate herself from being nailed down to this position. We often say "I dunno" when we think we do know but need to manage the contentiousness of the statement.

Another way of approaching the "how" is to be alert to figures of speech, particularly metaphors and metonymy.

Metonymy

This is where we conjure up an idea by connecting it to something that is associated with it. For instance, when we talk about "No. 10" or "the White House," we are talking about the British prime minister and the US president respectively. See Shank (2002: 166–167) for a useful discussion of the rhetorical function of metonymy.

Respondent Generated Metaphors

Whenever I code data, I always create a code for "metaphors" because they can provide telling clues as to how people see things. For instance, in the quote above, the respondent talks about his personality being "formed and moulded" through the successful living experiences and friendships he found. This molding metaphor conveys a sense of agency over self-development. Similarly, in another transcript from this research, a student comments that the "first hurdle is getting through my first year"—perhaps this conveys a perception of the undergraduate years as a series of obstacle courses. Another student talked about the "highs and lows" of university life, inviting us to think about the volatile nature of undergraduate experiences. Being alert to how people describe their experiences in this way supports a rich analysis and write-up.

Researcher Generated Metaphors

Another way of approaching metaphors is to derive them from the data yourself in relation to any variation you have identified. For instance, in her phenomenographic research into the nature of research, Brew (2001) suggested four metaphors (layering, domino, journey and trading) to capture how different academics described research activities in interviews she conducted with them.

Writing Up

There are a few points to be made about writing up and data analysis. The first is to resist treating the write-up as an end point in a linear process. In particular, your analytical memos are your rough draft of the report. The second point is the write-up is part of the analysis, as Coffey and Atkinson (1996: 109) put it, "writing makes us think about data in new and different ways." Writing is a sense-making activity in itself and, as such, it is not *about* the analysis, it is a deeper stage of it. Shank (2002) provides an excellent set of suggestions, guides and frameworks on writing up—see also Seale (1999).

You will need to make decisions about what you want to say about your own positionality. Surfacing researcher reflexivity in a brief set of notes is acceptable but writing indulgently about your own position as the researcher creates a text that is more about you than it is about the research topic. As Silverman (1997) has objected, you can overdo the "reflexivity card."

The best way of acquiring a writing style that suits both the research project and your own taste is to read other research reports to explore what is comfortable for you and what you would like to emulate. If you are happy with straddling the humanities and the social sciences, you may be drawn to Geertz's (1983) notion of qualitative research as a "blurred genre." Alternatively, you may prefer a more formal style. Above all, do not be boring. An important aim of qualitative research writing is to draw in the readers, to give them a vicarious experience of being "there" in the setting with you, talking and listening to the participants as you did.

Conclusion

Qualitative data analysis is a creative engagement with the empirical data, the literature and, where possible, conversations with the people who provided the data. Careful engagement with this combination supports our search for understandings and insights into human experiences, conditions and perceptions.

Further Reading

Coffey, A. and Atkinson, P. (1996) *Making sense of qualitative data*. Thousand Oaks, CA, and London: Sage.

Delamont, S. (2002) *Fieldwork in educational settings: Methods, pitfalls and perspectives*. London: Routledge.

Dick, B. (n.d.) *Grounded theory: A thumbnail sketch*. Available online at: www.scu.edu.au/schools/gcm/ar/arp/grounded.html#a_gt_intro (accessed 2008).

Glaser, B.G. and Strauss, A.L. (1967) *The discovery of grounded theory: Strategies for qualitative research*. Chicago, IL: Aldine Publishing Company.

Grbich, C. (2007) *Qualitative data analysis: An introduction*. London: Sage.

Shank, G.D. (2002) *Qualitative research: A personal skills approach*. Upper Saddle River, NJ: Merrill Prentice Hall.

Silverman, D. (1997) Introducing qualitative research. In D. Silverman (Ed.), *Qualitative research: Theory, method and practice* (pp. 1–7). London: Sage.

Walford, G. (2001) *Doing qualitative educational research: A personal guide to the research process*. London: Continuum.

4

FOCUS GROUP RESEARCH

Appeal

Focus group research appeals to many higher education researchers because its data gathering process extends the academic practice of exploratory discussion (in seminars, conferences, etc.). Focus group research is based on the principle that rich data can be elicited from group interactivity. In higher education research, focus group would support research into the following:

a) Students' experiences of higher education (formal and informal).
b) Supporting the formulation of survey or interview questions.
c) Evaluating programs, educational software, support services and so forth.
d) Adding evidence to findings from other methods of research.
e) Generating new ideas.
f) Problem setting and problem solving.
g) Exploring difficulties in the curriculum.

Purpose of Focus Group

Sharing and Comparing

Focus group researchers rely on the notion that when people gather to talk about something, their contributions and understandings will be enriched by the group dynamic. In Morgan's (1997: 21) words, this research method allows for "sharing and comparing":

Participants in focus groups often say the most interesting aspect of their discussions is the chance to "share and compare" their ideas and experiences. From the researcher's point of view, this process of sharing and comparing provides the rare opportunity to collect direct evidence on how the participants themselves understand their similarities and differences. This actual observation of consensus and diversity is something that can happen, quite powerfully through group interaction.

Co-Constructing Knowledge

The sharing and comparing of which Morgan speaks can also be seen as a developmental process which supports the focus group members to clarify, extend and review their understandings. This would be a constructionist interpretation of the purpose of a focus group. By this interpretation the research is a means by which knowledge can be co-constructed from a combination of the group discussion, the thoughtful prompts of the moderator and a theoretically engaged analysis (ideally shared with the focus group members). Within this framework, focus group research would be seen as generative of understandings rather than truth seeking.

Moderator not Interviewer

The use of the term "moderator" rather than "interviewer" for focus group research provides a reminder that the researcher role is to prompt and facilitate discussions rather than to control them.

The Uses of Focus Group Research

Focus group research is popular in the commercial world where it is often used to test new products, services and markets or to evaluate an organization's morale (Greenbaum, 1998). Increasingly, it is used in the social sciences as a means of exploring the experiences of particular groups including those who are considered to be hard to reach (e.g. refugees, drug takers).

Focus group research can be stand alone or it can sit alongside other methods. Also, as indicated, it can be the basis for exploring the best kind of questions to ask for survey research. Focus group research tends to gather convergent or common views and experiences. If you want to know what *most* students experience or think about an issue, it is a good method. That is not to say that the moderator cannot capture dissenting voices and diverse experiences within a group.

A Cheap Option?

Care has to be taken that focus group interviews are represented as *group* interviews. One focus group meeting with ten participants cannot be claimed to be "ten interviews." It is often suggested that focus group research is economical on time and money but this is not always the case because recruiting participants can be a headache. Making a decision to conduct focus group research on grounds of economy undermines its singular purpose to secure data through social interactivity.

And Then We Will Do Focus Group Research . . .

In my experience, too many people dive into focus group research as if its demands were self-evident. I have seen many research designs which plump for focus group research as a catch-all way of handling the qualitative side of data collection. First, there is a proposal to do the serious quantitative work and next there is a promise to get some people in a room to verify the findings. Focus group research can certainly complement quantitative research but it is an alternative source of intelligence rather than a means by which statistics are prettified.

Theoretical Concerns

Trustworthiness

Trustworthiness for focus group researchers tends to rest on two key moves, namely purposive sampling and—from grounded theory—achieving saturation (Glaser and Strauss, 1967). The first guides the researcher to think about what kind of comparisons across groups he or she is going to make. Often this means considering what kind of diversity each of the groups need to reflect to ensure a range of views are gathered about a common experience. For instance, the participants may all be undergraduates but the researcher may need to explore his or her topics with diverse cohorts or mixes of undergraduates (e.g. home students, international students, first year, arts, humanities). Saturation requires that the researcher continues to explore particular lines of inquiry with further groups until she or he is satisfied that no new insights are emerging.

Other moves to strengthen the trustworthiness of this research method include having an observer to watch and comment upon the discussion process, moderator reflexivity (captured in field notes) and sharing the transcript analysis with the participants and/or other researchers. I will pick up these issues again under Method.

Group Dynamics

Because focus group research relies on the sharing and comparing process, the moderator has to have some familiarity with theories of group dynamics. Stewart and Shamdasani (1990) define three areas to help the moderator anticipate what might emerge from the group, namely the interpersonal, the intrapersonal and the environmental.

Interpersonal

This refers to both the relations among the focus group participants and with the moderator. Much research has been dedicated to ways in which power and a diversity of communication styles circulate any room full of people. The particular power issues to watch for in focus group research connect to age, sex and status.

Age Too much age differential in the room will reduce the quality of the discussion and researchers are advised to go for similar age groups. This might be about power more than age. In my own experience, where a common identity overrides age as in the case of a group of subject specialists, it did not seem to be a divisive factor.

Gender   It is common to see women and men as having different "interactional styles" (Stewart and Shamdasani, 1990: 143), with the latter tending to dominate group discussion though these must be treated as tendencies that do not always occur. Rather than prejudge the situation by avoiding mixed sex groups, it is best to see what happens with the groups you recruit and adjust your strategy accordingly. In my experience of mixed group discussion, there has not always been a gendered pattern of interaction which privileges one sex. Clearly, decisions have to be made about the purpose of the inquiry and the extent to which same sex or mixed sex groups will suit the aim. For instance, in British universities male students are not achieving as well as female students (except at the very top) and an inquiry into this would suggest an exploration with single sex groups.

Ethnicity   Ethnicity is another variable that researchers are fond of manipulating. It is extremely important not to overdetermine participants or the moderator as members of an ethnic minority group as if this factor always shapes what they will say or experience. Equally, it is important to have a sense of when ethnicity does matter and to duly engage with how this affects group interactivity. A tricky path has to be navigated by moderators in order to avoid the twin risks of stereotypically overstating difference or of ignoring it.

Social power   Social power is often based on status and most researchers try to avoid disparities of this kind in the groups they recruit because it is known to affect disclosure. Other key forms of social differentiation concern wealth and expertise.

Intrapersonal

This refers to what everybody brings into the room in the way of their unique personal biography and dispositions. It is this dimension which threatens our attempts to make predictions about the behavior of particular groups and individuals because we are always more than our social or biological category. Stewart and Shamdasani (1990: 39) offer two broad orientations on group participant disposition which might help the moderator to assess the mix she or he has in the room.

Social sensitivity and *ascendant tendencies*   According to Stewart and Shamdasani (1990), those with a social sensitivity disposition tend to be responsive and attentive listeners. In contrast, those with ascendant tendencies tend to assume a more assertive and dominant role in groups. Doubtless there are participants in between these types but Stewart and Shamdasani (1990) have given us useful polar positions for thinking about the quality of interaction in the room. The moderator who is mindful of these positions, will be better equipped to make careful judgments about the facilitator moves required to ensure a balanced discussion.

In addition to team playing dispositions, there will be issues of self-management. What we say connects to how we want to project ourselves or what we think is appropriate. We may not publicly identify with certain positions that we hold privately. We may not disclose certain feelings or views in the presence of particular people. For instance, it is not usually a good idea for academic teachers to moderate a focus group discussion comprised of their own students. Basically, the moderator will need to be aware that to some degree or another, we all speak from defended positions.

Environmental

This refers to the influence of the setting on the group dynamic and ranges from concerns about the actual interior (e.g. whether it includes distracting posters on the wall or uncomfortable chairs) to the layout and space allowed between seated participants (e.g. does this compromise people's sense of safety and comfort?).

Convergence or Conformity

Further considerations about group behavior can be drawn from Potter and Weatherall's (1998) notion of interpretive repertoires and Foss' (1996) notion of rhetorical communities. These help us think about the need to use the focus group discussion as a reflexive one for the participants. In attempting to capture convergent views and experiences, it is important to avoid encouraging unproblematized narratives. The process of sharing and comparing should enable participants to progress their understandings.

Rhetorical communities   There is the ever present danger of reading the *construction of* convergence as an expression of some form of unmediated unanimity of opinion. Foss (1996: 125–126) has argued that when a group gets together, they are liable to develop forms of group talk in which members "will share the same vision of what counts as evidence, how to build a case and how to refute an argument." In addressing this, a moderator will need to consider whether some members of the group have set the moral tone for the discussion, or whether the mood set by the interactivity allows for dissenting voices.

Convergence is not always going to be a problem. Indeed for some research questions, the moderator will be interested precisely in the "group talk" as with, say, an exploration into how a disciplinary community sees itself. If, in contrast, the researchers want to explore how members of this community see the relationship between research and teaching in their subject, they will want to invite a more developmental thrust to the discussion within a spirit of co-inquiry.

Interpretive repertoires   Another way of conceptualizing group talk is through the concept of an interpretive repertoire. This concept is underpinned by the social constructionist position that the thinkable is constituted by the explanatory vocabulary and discourses available to us. According to Potter and Weatherall (1998), our discussions will draw on an available repertoire that will delineate how we see the problem. Characteristically, we think within the box and a focus group discussion should expand interpretive repertoires rather than be restricted by them. If it is compatible with the research question, focus group research can help to stretch the thinkable and to introduce new ways of talking about things.

Giving Voice

A further way of looking at group centered research is to grasp its benefits for marginalized groups because it may provide members with a more comfortable and safe setting for them to explore sensitive questions.

In her work with Latino women, Madriz (2000: 836) writes that focus group research:

... may facilitate women of color "writing culture together" by exposing not only the layers of oppression that have suppressed these women's expressions, but the forms of resistance that they use every day to deal with such oppressions ... focus group research can be an important element in the advancement of an agenda of social justice for women.

To illustrate, Madriz (2000: 835) offers revealing comments from one of her participants:

... when I am alone with an interviewer, I feel intimidated, scared. And if they call over the telephone, I never answer questions. How can I know what they really want or who they are.

Madriz (2000: 836) argues that the interactivity of the group, reduces the influence of the moderator. This is not an automatic effect of focus group research arrangements. Firstly, where the moderator has more status power than the participants, they may look to her rather than across to peers in responding to questions; secondly, the group interactivity might revolve around a powerful member of the group. This is one of the reasons why proponents of focus group research insist that the moderator needs to have an understanding of group dynamics.

Experiential affinity   Madriz (2000) suggests that the moderator needs to have some experiences in common with the group to establish trust and empathy. We often have to make compromises with this principle and indeed its significance depends upon the purpose of the research. It is important not to be too formulaic about diversity issues. There really are some topics such as racism which require sensitive attention to a suitable moderator, while others can be approached more flexibly. Over the years, I have conducted focus group research among a number of student groups and though aware of my own position as an older academic researcher, I have managed to gather interesting and useful data. A student moderating the same groups may well have got different kinds of data though hopefully some of it would have overlapped with mine.

Online Focus Group

Frances Deepwell and myself (Cousin et al., 2005) experimented with a combination of online and face-to-face focus group research for product evaluation. Academic colleagues were invited to test a selection of commercial software for electronic learning environments and to take part in an asynchronous focus group discussion about them. We then met the group members face to face, presented them with their views so far and got them to vote on which software they preferred. This was a relatively simple use of the online medium and since our tentative steps, the academic world has grown much more used to e-discussion forum, emoticons and asynchronicity.

There has been a growing realization that contributing to discussions online does not necessarily constitute a disadvantage with respect to face to face research. Asynchronicity, Joinson (2005: 32, 33) writes:

> Reduces the cognitive load associated with the need to combine answering a question with impression management . . . producing better quality data, particularly when dealing with sensitive topics.

Like Joinson (2005), a number of researchers (Hine, 2005) have found that some people feel less vulnerable and exposed by participating in virtual discussion rather than real. Another advantage of conducting focus groups online is that contributions are already typed and ready to be analyzed.

Method

Recruitment of Participants

Often access to focus group participants is the biggest problem the researcher faces. Sometimes there are difficulties in getting a group to agree a time, place and date or of not turning up on the day. Offering book tokens and refreshments has become standard practice, particularly for recruiting students to focus group research. If appropriate for the research, you can poach time for a focus group meeting from an existing event (e.g. departmental meeting or seminar).

Single Group

Some forms of focus group research are confined to a single meeting because the main purpose is idea generation. Alternatively, a series of focus group meetings can be held with the same cohort if the goal is to explore an initiative over time or to evaluate a project.

Size and Duration of Focus Group Meeting

A focus group usually involves about a minimum of four and a maximum of twelve people; each session lasts between one to two hours, sometimes with a scheduled break.

Segmentation and Purposive Sampling

Most researchers recommend recruiting fairly homogenous groups of people who do not know each other to maximize the expression of a diversity of views. I have not always worked with this principle and have recruited from pre-existing groups of academic colleagues or students. From this experience, I am not convinced that it is necessary for the group members to be strangers to each other.

Participants need to have something in common (e.g. biology undergraduates) for the purposes of group cohesion and good discussion. Morgan (1997: 36) suggests that if you think the group could easily discuss the topic in everyday settings, they are likely to be good participants. Choosing a group with characteristics in common is known as segmentation. In choosing how many focus groups to run, you will want to consider the diversity of experiences that need to be reflected for each separate focus group meeting (e.g. age, subject, etc.) in order to make comparisons (purposive sampling (Barbour 2007: 58)). Put simply, a researcher might convene three or four focus groups from a segment of the population that has something in common (all undergraduates). Each successive group meeting might contain variation within this segment (male/female, etc.) so that views across the groups can be compared.

Saturation

While you may plan to do a set number of focus groups (say, four), you can make decisions about whether you need to do more on the basis of the principle of saturation. All this means is that if your data is generating the same things, it is saturated. Thus if four groups of biology students tell you that the mathematics component of their first year program is difficult, then you can conclude fairly safely that this is so. If you are getting inconsistent data about the difficulty of the subject, you may want to consider convening more groups. But do not slavishly adhere to the principle of saturation, particularly if the subsequent groups do not appear to be building on previous ones (this is unlikely). This could mean that there is not a consistent experience to capture. You might switch to a concern for the complexity of the experience rather than convergent testimonies of it. Or you might have generated insights from the group discussions that went exceptionally well and decide to focus on these.

Decide the Focus

At the risk of stating the obvious, focus group research requires a focus. The design begins with a relevant research question that expresses this focus and that can inform the construction of the prompt sheet. Here are some examples from focus groups I have run with academics and students in UK universities:

a) Are there any concepts that stand out as difficult or troublesome in this subject?
b) How do international students experience studying in this UK university?
c) Are there conceptual and/or practical links between teaching and research in architecture, town planning and urban geography?
d) What are the experiences of reading and writing for civil engineering students?

Reading the Literature

As with all qualitative research, focus group research requires an engagement with the relevant literature to:

1. Shape the research focus.
2. Inform the data gathering and the analysis.
3. Support researcher reflexivity and theorizing.

The generation of understandings and insights from focus group research comes from a constant engagement with the empirical data, the literature and the moderator's reflexivity. These are not sequential activities in a research cycle that starts with the reading and ends with the analysis and write-up.

Formulating Prompts: Low/High Moderation

You will need to prepare a prompt sheet of questions for flexible use in the discussion. These questions will be organized around the topics you want to cover; you may find that you will need to reformulate or drop some of the questions once the discussion gets going. If it is feasible, it is a good idea to pre-test questions or at least to show a draft to other researchers for their comments. Morgan (1997) distinguishes between low moderation where about two questions are prepared and high moderation where around five or six questions will be prepared. If the participants feel strongly about the topic and if there are a range of views in the room, it will take very few questions to get the desired interactivity and discussion (Stewart and Shamdasani, 1990: 62).

The following are examples of "high moderation" questions to explore the international student experience on a particular campus. The bracketed points serve as prompts to use sparingly in case the discussion dries up or important areas are not covered spontaneously. The "further" in brackets acknowledge that the members may have already said something about this but you want to probe further.

- Can you take me through your first experiences in the first six months at Poppleton? (Are the experiences shared, which experiences stood out? Any surprises? Shocks?)
- Can you comment (further) on the social side of your stay here? (Relations with home students? Other international students? Host families? Local community?)
- Can you say something (more) about the teaching at Poppleton? (Anything unexpected? Strange? How different to back home?)
- Can you comment (further) on the academic requirements here? (Assignments, deadlines, essay writing in English, group work, plagiarism.)
- Can you say something (further) about the support you get for your studies? (English language, peer support, study skills, personal tutor.)
- What advice would you give an international student coming to Poppleton?

Once you try out these kind of questions and prompts, you will want to review, cut or combine some, depending on which work well to develop the group discussion. I have started with a "grand tour" question because it is a classic interview opener which establishes the direction of travel for the discussion in fairly unthreatening ways. With the exception of the last "closing" question, they are all framed as invitations for a conversation indicated by the "can you comment/ say" beginnings. You will aim to get to the heart of the matter under discussion when the group energy and interest is high. Morgan (1997: 51) suggests that this is around ten minutes into the discussion.

Low Moderation Question

Morgan (1997: 18) asked one group of widows "What kind of things have made being widowed either easier for you or harder for you?" and left them to discuss this with "only minimal guidance" from himself as a moderator. It was his only prepared question (though it is really two questions). Clearly judgments have to be made about whether low moderation of this sort would work with the groups you convene.

Moderating

The quality of data is going to have much to do with the moderator's skills. Firstly I will deal with a few challenges he or she faces and then some strategies for encouraging good quality interactions.

Tolerating Silences

Most experts advise the moderator to tolerate silence a little longer than might be comfortable to encourage the group to break the silence. Remember, the aim is for the participants to control the group dynamic as much as possible and this would be an important test.

Non-Verbal Communication

Stewart and Shamdasani (1990) urge that attention be paid to non-verbal cues, recommending that focus groups be videotaped or watched from behind a double mirror. From an ethical viewpoint (let alone the practical), I do not think that many academics would want to bring double mirrors into the field of higher education research. However, videotaping is a realistic option for some contexts. If you are limited to the tape, it will pick up laughter, groans and silences—all of which can be included in the transcript. Finally, a moderator might want to note significant verbal cues as the discussion proceeds.

Moderator Bias

Moderator bias can operate in a number of ways, namely: favoring the contributions of particular participants, directing positive body language towards some participants, indicating agreement to some and not others, supporting the most animated of the group and neglecting the quieter ones and so forth (Stewart and Shamdasani, 1990).

Moderator Skills

There is no shortage of advice in the literature on moderation strategies and skills that support the interactive dynamic. In drawing on the issues discussed under "theoretical concerns" above, there are two main things to stress: firstly, that you are not chairing a meeting or conducting an interview. You are encouraging the participants to talk and anything that threatens the group dynamic for this (e.g. dominant speakers, too many questions, cold room, self-appointed experts, silent participants, aggression) needs sensitive handling. Secondly, if participants offer their opinions, try to get them to talk specifically about any experiences that relate to them. If a male student says that he thinks the assignment marking is unfair in his department, the data is clearly going to be richer if you can get him to support this view with an experience. You will then be able to ask other participants if they have shared experiences.

Ground Rules and Warm Up

The moderator needs to invest some initial time in communicating the ethical framework of the event. Make this section snappy so as not to lose the interest of the group with too many preliminaries. Something needs to be said about confidentiality (if this is relevant) and the right to leave the discussion or to not be quoted. The moderator communicates an expectation that no-one interrupts others and that divergent views are respected. The moderator needs to indicate that the event will be videotaped (having tested the equipment).

The moderator also indicates that towards the end of the session, he or she will attempt to summarise key points so that participants can correct, add or expand on these.

Observer

Some focus group meetings have an observer and clearly his or her presence needs to be explained. Other meetings have a note taker to pick up non-verbal cues, names of those speaking and to support the summary. Neither of these are necessary but if I could choose just one, I would go for the observer because his or her reflections on the process are additional data.

Identifying Participants

A useful first warm up question asks each of the group to introduce themselves and to offer some basic information. This can be requested on a short form so that they can be handed in to aid transcription and analysis. Where possible, the moderator uses participants' names in discussion to aid the transcriber in identifying individual voices. It is helpful to provide participants with folded paper so that they can display their names to each other and to the moderator.

The following are some interesting strategies for getting groups talking.

Stimulus Material

Groups can be asked to look at prepared material to stimulate discussion: this can take the form of a case study, a newspaper article, a hypothetical case, a video clip, etc.

Writing

Barbour (2007) suggests that one way of ensuring that participants are not overly influenced by groupthink is to give them a brief writing exercise to secure their own thoughts first. These written pieces can become part of the data.

Prompting Disclosure

One way of getting a group to open up is to report one or two of the things a previous group has said and to invite comment on this.

Metaphors

Towards the end of discussions, I sometimes ask participants to think of a metaphor that describes the problem/topic/project under discussion. This can turn out to provide rich, evocative data though you have to tolerate quite a long silence while participants think about this.

Word Association

Where the topic under discussion is sensitive, I have invited participants to word associate. For instance, I asked a group of civil engineer students with whom I was exploring writing (which they disliked doing) to give me the first thing that entered their head against the words "grammar and spelling." In return I got: "nightmare," "infant school," "primary school," "horror," "fail," "difficult." This led to discussion about the emotional dimension of writing. Were I to have asked "how do you feel about writing," the responses may not have been so forthcoming (Davies and Cousin, 2002).

Visual Representation

Together with another colleague, I once ran a focus group on "feedback" with students by providing many magazines, scissors and glue and asking them to visually represent the issue. Once the students (in groups of three) had discussed and completed their "poster," they used them as a basis for full group discussion. We were then able to explore the views of academic colleagues by showing them the student posters as stimulus material.

Role Play

The group can be asked to begin the session with a role play exercise that can then form the basis for the discussion. Many people are uncomfortable with role play and the moderator needs to be confident that it is both appropriate and feasible.

Provisional Co-Analysis

Barbour (2007: 80) advises that the moderator works with the participants to share provisional analyses. The "good moderator" she writes "should also keep a weather eye open for distinctions, qualifications and tensions that have analytic promise."

This sharing of emergent conceptual understandings with participants will help them to think deeply about the topics in hand. This involves offering a provisional statement such as:

> It sounds like you are always defined as international students rather
> than as merely students. Does that sound right?

Clearly, this kind of comment would need to be made sparingly
and very tentatively to avoid over-suggestion by the moderator or
premature closure of the discussion.

Summarizing and Ranking the Key Points

As indicated, towards the end of the process, you attempt to play
back to the group what you think they have said of importance.
Allow a little time for them to comment on this. Some moderators
ask the members to rank factors in terms of importance to get a
rough handle on what seems to matter most to the group.

Field Notes

Once you have finished the focus group research and thanked the
participants, you will need to write up your own reflections of the
event. If you have an observer, you will want to hear his or her
reflections to add to yours.

Data Analysis

Returning to the Field

You might feel that you do not have enough data in an area once
you come to analyze your data. Perhaps you realize that you did not
prompt enough discussion about a particular area. In this case, you
can recall earlier participants to explore this or you can convene a
"wildcard" group (Barbour, 2007: 65) of new participants. Another
option is to interview individually a handful of participants. Or you
could email the participants your thoughts on this area, asking them
to add their comments.

Unit of Analysis

The primary unit of analysis is the group but attention can also be
paid to individual contributions. This means that data chunks will

vary from a set of exchanges to show argument (or conflict) development to single, stand alone contributions by individual participants.

What is Emerging

Barbour (2007: 131) suggests that the first stage of analysis concerns the identification and depiction of patterns. To this end you devise a grid which "allows you to see at a glance the preponderance and distribution of comments on particular themes in the various groups." Exploring patterns in this way is a good starting point so long as we bear in mind that some of the insights you generate will come from single contributions or indeed from what is not said. Accompany this patterned behavior with notes about what does not fit into the pattern and any surprising, noteworthy absences.

Every transcription of each focus group meeting will constitute a data set (to include your reflective notes). You might start your analysis with a grounded theory approach (Glaser and Strauss, 1967) but see Chapter 3 for more on this approach and further ideas as to how to approach the analysis.

Conclusion

Many focus group meetings are convened and conducted without regard to their underpinning theory and a methodological approach. This form of research provides data out of the dynamic of group discussion and behavior—this needs to be appreciated and pursued to get the best out of this versatile approach.

Further Reading

Barbour, R. (2007) *Doing focus groups*. London: Sage.
Morgan, D.L. (1997) *Focus groups as qualitative research*. London: Sage.
Stewart, D.W. and Shamdasani, P.N. (1990) *Focus groups: Theory and practice*. London: Sage.

5

SEMI-STRUCTURED INTERVIEWS

This chapter focusses on semi-structured interviews because they are the most adopted form of interview in qualitative research.

Appeal

Semi-structured interviews allow researchers to develop in-depth accounts of experiences and perceptions with individuals. By collecting and transcribing interview talk, the researcher can produce rich empirical data about the lives and perspectives of individuals.

Purposes

Type of Interview

There are three kinds of interview: structured, semi-structured and unstructured. Structured interviews are essentially face-to-face surveys where mainly closed questions are asked (online, telephone or live) against coded responses. Unstructured interviews are where the researcher guides naturally occurring conversations. The line between unstructured and semi-structured is fuzzy because, as Gilham puts it (2000: 3), "expert interviewers always have a structure, which they use flexibly according to what emerges." More on unstructured interviewing can be found in Spradley's (1979) excellent book *The Ethnographic Interview*, discussed in Chapter 7 (Ethnographic Approaches).

Semi-structured interviews are so-called because the interview is structured around a set of themes which serve as a guide to facilitate interview talk. Unlike the structured interview, the interviewer is expected to adapt, modify and add to the prepared questions if the flow of the interview talk suggests it.

Researching Complexity

Semi-structured and unstructured interviews attempt to grapple with complex experiences. If you need to gather relatively straightforward information, such as whether students live at home or in residences, a survey will do the job. If, on the other hand, you want to know whether particular groups of students feel "at home" on campus, you will probably need to talk to them either as a group (see chapter on focus group research) or in one-to-one in depth interviews so that you can get at more layers of meaning. Rubin and Rubin (2005: 2) offer the following illustration of this point:

> You could measure poverty by counting the dollar income people earn, before taxes. But you would learn little about how people survive on their income, whether they share apartments with boarders, whether they fix the plumber's car in exchange for plumbing services, or whether they eat at relatives' houses at the end of the month when they run out of money.

Some research designs will include both the equivalent of "counting the dollar income" to get a broad, contextual picture and in depth interviews to drill down into how people operate within this context. In a number of cases, new data will not need to be generated at the broad level because it already exists. For instance, in the UK, the Higher Education Statistical Agency (HESA) routinely gathers data on recruitment and attainment patterns broken down by ethnicity, gender, university, subject, etc. There are equivalent agencies in other countries. What is often needed in higher education is research which offers meanings to the patterns and variation found in statistical evidence. Semi-structured interviews with relevant groups and individuals is a good way of searching for such meanings.

Theoretical Concerns

Interviews and Neutrality

On the face of it, interviewing seems a relatively straightforward stimulus-response event. You ask the questions and the respondent responds but they are far from a neutral information eliciting tool. Fontana and Frey (2000: 645) point out that "asking questions and getting answers is a much harder task than it may seem at first. The spoken or written word has always a residue of ambiguity, no matter how carefully we word the questions and how carefully we code and report the answers." Gathering and representing people's experiences is fraught with interpretive difficulties.

We should also problematize the relational character of interviews. As Schostack (2006: 1) writes, "the interview is not a simple tool with which to mine information. It is a place where views may clash, deceive, seduce, enchant." Similarly, Holstein and Gubrium (1997: 116) object to the "image of the social scientific prospector" which "casts the interview as a search-and-discovery mission" rather than an interactional event in which meaning-making is *in situ* and the product of both players in the interview rather than that of the skilful transcript analyst after the event. The interview, write Alldred and Gilles (2002: 146) "is the joint production of an account by interviewer and interviewee" and again, to quote, Holstein and Gubrium (1997: 14) "meaning is not merely elicited by apt questioning nor simply transported through respondent replies; it is actively and communicatively assembled in the interview encounter."

Interviews as Meaning Making Events

Abandoning the idea of the interview as a straightforward site of excavation means acknowledging its meaning making character. Perhaps in this light, unstructured or semi-structured interviews are best conceptualized as a "third space" where interviewer and interviewee work together to develop understandings. Davies (1999: 96, 97) gives a very good illustration of this dynamic in drawing on her research with British parents of a young man with learning difficulties. At the beginning of the interview the parents declare that their son has "got no value of money." Thirteen responses later, the father concedes: "yes probably he would value a bit of money, if he was having it in a pay packet every week." Good interviews, then, allow a dialogic, reflective journey to take place between interviewer and interviewee.

If you see the interview as a kind of third space, then you might be drawn to Holstein and Gubrium's (1997) model of "active interviewing" in which interviewers offer interpretations, connections, ideas and possible conceptual hooks to support an explicit, dialogic meaning-making direction for the interview. This shifts the method talk for interviewing from the formulation of the right questions for the right responses to the formulation of moves to develop the dialogue and analytic attention to how meanings are assembled.

In their extremely helpful chapter Active Interviewing, Holstein and Gubrium (1997) argue for there to be dual attention to both the *how* and the *what* of an interview. This requires an interest in both the content of the interview and in the ways in which such content is assembled. Holstein and Gubrium (1997) take the reader through an example of a woman who is caring for her elderly mother. This woman is seen to use various "narrative resources" as wife, mother and daughter to present her predicament. Quite simply, what she says is inextricably linked to how she positions herself at various points in her account. Rather than simply capture this linkage, the aim of the active interviewer is to think *with* the interviewees in order to extend understandings from this linkage.

Horizons of Meanings

The aim of an active interview, then, is to access and expand on the interviewee's understandings, however tentative or contradictory these may be. This often involves suggesting "possible horizons of meaning" —it is worth quoting Holstein and Gubrium (1997: 125) in full on this:

> The active interviewer sets the general parameters for responses, con-
> straining as well as provoking answers that are germane to the researcher's
> interest. He or she does not tell the respondents what to say, but offers
> them pertinent ways of conceptualizing issues and making connections—

that is, suggests possible horizons of meaning and narrative linkages that coalesce into the emerging responses.

This orientation on the interview might be seen as one in which interviewers and interviewees alternate as teachers and learners throughout the interview process (Pawson and Tilley, 1997). Sometimes the interviewers need to suggest explanations as teachers and sometimes they need to listen and hear explanations as students.

The Art of Hearing Data

Above all, the best kind of interviewer is said to know how to listen. This is well expressed in Rubin and Rubin's (2005) model of "responsive interviewing" which they express as "the art of hearing data." Those supporting an "active interviewing" model might see an emphasis on hearing data as an emphasis on the "what" at the expense of the "how." Equally, however, the image of hearing as an art reminds us that if the interviewer jumps in too many times, he or she may end up dominating the event.

The Social Position of the Researcher

The researcher must listen out for how the "circumstances of the interview" (Atkinson and Hammersley, 1994: 113) might be shaping what the interviewee is saying. In short, are there consequential power asymmetries between interviewer and interviewee? Anne Oakley (1990) was among the first to draw our attention to the power imbalances between interviewee and interviewer and Holstein and Gubrium's (2003) edited book offers a more recent discussion of this issue. Power imbalances may be to do with gender, rank, disability, class, ethnicity as well as the asymmetry between discloser and disclosee.

Broadly, there are three ways of addressing researcher positionality. Firstly, where it seems a good move, efforts can be made to establish some form of commonality between the interviewer and the interviewee. For instance, students can be asked to interview students. Do not be too formulaic about this principle, particularly in the assumption that common markers of identity (e.g. female or black) offer automatic sources of empathy. This depends on the capabilities of the individuals concerned as well as on their shared identity position. Further, sometimes differences *enable* conversations (Miller and Glassner, 1997). Who interviews who is a practical and situated matter.

Secondly, the interviewer needs to do his/her best to minimize the power present in the interview by, for instance, disclosing their own relevant experiences and by facilitating an exploratory thrust rather than an information prospecting one.

Thirdly, the researcher needs to take a reflexive stance by problematizing positionality throughout the interview process. More generally, it is important that the interviewer develops a conversational style, building a trust relationship with the interviewee. The more distance between interviewer and interviewee, the less trustworthy the responses are likely to be. This intimately connects to the next issue.

Building Rapport

Most researchers place importance on the building of rapport between interviewer and interviewee. This reverses the positivist position that requires a researcher stance of neutrality and objectivity, particularly since one of the moves for rapport building involves measures of disclosure from the interviewer. For most interviewers, the quality of the conversation between the interviewer and the interviewee is in large measure dependent on the rapport building capacity of the researcher.

Embellishing Responses or Pleasing Behavior

Most people tend to place themselves in a good light in the telling of stories and the offering of opinions. How can the interviewer be sure that this tendency has not inflected the interview? Similarly, if the interviewee knows that you are seeking out bad news stories, how can you be sure that they are not obligingly providing them at the cost of a more balanced report of experiences? An exploratory, dialogic direction to the interview might guard against inviting an interviewer-pleasing or embellishing posture from the interviewee.

Giving Voice

Interviewers are often struck by the readiness with which interviewees will talk to them. Eisner (1991: 218) offers an explanation:

> Researchers can take advantage of a quasi-therapeutic relationship because of the attractiveness of one of our most treasured gifts to others— the gift of lending serious attention and a sympathetic ear to what someone has to tell.

Bearing this in mind, the interviewer needs to minimize the risk of straying into a therapeutic domain without the skills or licence of a therapist. Again, the model of active interviewing offered by Holstein and Gubrium (1997) provides a way of avoiding this risk because of the centrality it gives to a developmental dialogue.

More vulnerable to a therapeutic direction is research which aspires to "capture" the voices of the marginalized. Sometimes this aspiration results in the production of a collection of victim narratives which appear to have been prompted by an interviewer taking a quasi therapeutic stance. Do not enter the interview stage with a deficit model explanation of your interviewees or a political investment in over-determining them as different and "other." We are always more than the categories assigned to us and an interview should not be in the service of a narrow hunt for social difference data. The point of an interview's indepthness is to explore subtleties, nuances, uniqueness and singularity alongside possible generalities and commonalities across groups.

Non-Linguistic Communication

Non-linguistic communication will impact on the interview: this includes interpersonal space, use of pauses and silences, tone of voice, body language, dress and other modes of presentation of self from the interviewer (Fontana and Frey, 2000: 660–661). The interview setting will also convey messages to be addressed: is it comfortable, appropriate, easy to access and so forth? If you are aiming for an alternating teacher–student dynamic, is the environment friendly to this?

Epistemological Position

In order to accommodate the idea that the interview is a site of meaning making and "data making" (Baker, 1997) as much as it is a data gathering one, I suggest adopting Miller and Glassner's (1997) interactionist position which is mindful of but outside of the "objectivist-constructionist" continuum. Miller and Glassner (1997: 99) write: "we are unwilling to discount entirely the possibility of learning about the social world beyond the interview in our analyses of interview data." Their perspective accepts that interviews make meanings *in situ* but it also defends researcher authority in the shaping of the interview and the analyzing of the data to understand something of the social world. Miller and Glassner (1997) offer a path between the view that there is an entirely knowable, objective world "out there" (objectivist) and one that says that we cannot know anything "out there" because reality is primarily "in here" and the stuff of highly situated and negotiated meanings (radical constructionists).

Ethical Framework

An interview "requires that the interviewee consents to the interviewer to ask him or her questions on an agreed topic and to use the resulting transcript for research purposes." This blunt delineation of the interview from Denscombe (2007: 173–174) is helpful in thinking about the ethical framework. To accommodate the discussion above, you might want to reframe this as "An interview requires that the interviewee consents to a sustained dialogue on an agreed topic and consents to the interviewer to use the resulting transcript."

Informed consent often needs to be ongoing rather than a preliminary stage of the interviewing; this includes letting the interviewee see the transcript and comment on your analysis if you can.

Trustworthiness

Most of the above points have a bearing on the question of trustworthiness.

Here are some of the things you might bear in mind to ensure a trustworthy research report:

- Have you avoided smoothing your interpretation? For instance, have you been sensitive to variation of experience and viewpoint within groups as well as across them?
- What has determined your decision about numbers of interviews? Can you defend this?
- Have you shared your emerging ideas with the interviewees throughout the interview?
- Did you build rapport with the interviewee?
- Have you asked a colleague to check some of your transcripts/tapes to explore rival explanations and interpretations?
- Have you kept a diary to capture your reflections, theoretical leads and so forth?
- Does your research report avoid cherry picking quotes?
- Have you displayed your own questions as well as responses to show a developmental dialogue?
- Have you explored alternative data sources to strengthen your interpretations?

Method

Sample

In choosing the spread and numbers of interviewees, qualitative researchers often rely on "purposive sampling." This simply means recruiting people on the basis of a shared characteristic which will help you in your inquiry. For instance, if you want to know how students experience their first semester at university, you might want to identify around six to ten first year students as well as the same number of final year students. The first cluster will have strong, immediate recall and the second can be invited to reflect back to their first semester from the benefit of a longer undergraduate experience with which to compare it. You can try also to get a cross section of other kinds of experiences such as students living away on campus and students living at home, male and female, etc. By choosing your groups, you are not trying to exhaust any possible variation—this is not the goal of qualitative research. You are selecting what you think are key sources of variation to add to the depth and plausibility of your analysis.

Theoretical Sampling

If you decide to go for the grounded theory (Glaser and Strauss, 1967) strategy of theoretical sampling, this will allow you to be open about how many interviewees you recruit. Theoretical sampling involves two moves. Firstly, you must analyze your data as you interview (good advice for any qualitative research) and second, you begin to make judgments on the basis of the emerging analysis about how many more people you want to interview and what kind of experiences you would like them to reflect. Theoretical sampling means that you talk to more people to go more deeply into the issue, not to accomplish coverage of all possible sources of variation. The notion of saturation, also from grounded theory (Glaser and Strauss, 1967), will come to your aid in judging the number of interviews to conduct.

Saturation

As the term suggests, you have achieved saturation of a theme or finding if your fresh interviews are not telling you anything new. Once you have decided that you have reached saturation from a particular line of inquiry, it makes sense to stop interviewing about it. The challenge will be to avoid premature closure in making this decision.

Setting and Process

Some thought needs to go into agreeing a comfortable setting for the interview. Let the interviewee know that you are hoping for an event that will not be overly formal. Explain that you want to interview them because they have the expertise and experience you are seeking to learn from. And, of course, give them some indication of the purpose and ethical framework of the research. I usually give the interviewee brief typed information about the purpose of the research project, the confidentiality of the interview and my contact details. I secure consent for taping the event and for using the eventual transcript (duly anonymized), letting the interviewee know that I will send it to him or her to check and to add any further comments (much of this will need to be repeated at the beginning of the interview).

Research Focus

Because of their open, exploratory nature, in depth interviews are not well suited to hypothesis driven research (Rubin and Rubin, 2005: 40). They are more suited to broad "what is going on here?" type questions or to responding to a puzzle such as "why do students choose co-nationals for group work." They can also be used to support an exploration into identity formation (see Chapter 6 on narrative inquiry). As I discuss below, the trialling process will help to refine the question and the focus; it will also support the devising of promising interview questions and the discarding of others that needlessly clutter the interview.

Ask a Stupid Question . . .

Here is Gillham's (2000: 21) advice on how to design a bad interview:

> The way to construct a disastrous interview or questionnaire is just to sit down and knock out a set of questions off the top of your head. As anyone with any experience of higher education will testify, this is all too often how it is done by students at many levels. The resulting data are not only poor but often virtually impossible to analyse.

Invariably, most of these "knocked out" questions will be centered directly on what the researcher is looking for because this seems the most obvious line of inquiry. But if you want to know, say, if a group of students are tempted to plagiarize, the least helpful questions will pointedly ask why and when they have succumbed to plagiarism. You are better off formulating a few questions that get them talking about how they study for an assignment and their views of the assessment requirements of the university. This acknowledges that people do not always have a consciously worked out reason for why they do certain things. If the interview aims to be developmental, you need to come up with questions that are more than information seeking.

Trialling

Coming up with facilitative questions is a matter of trial, error and practice. Gillham (2000: 20) insists that the only way to get good at devising interview questions is to take the trialling phase seriously:

> No matter how well you have thought through the topics or questions, the actual running of them will be a chastening experience. This is reality, not something you have worked out on paper or in the comfort of your own head.
>
> (Gillham, 2000: 23)

The trialling process enables the development, grouping and timing of the themes. A face-to-face interview should be around an hour and a telephone interview forty-five minutes. Generally, follow a principle of "less is more" in determining the number of themes you want to cover. This might involve refining your research question in the light of the responses you have received. Once you have a sense of direction from the trialling process, you can firm up your research question and formulate the interview schedule. Gillham (2000) stresses that trialling is about exploring good questions and that it is not a substitute for piloting which must follow the trial process.

Trialling, Piloting and Active Interviews

I am not sure how far you can apply the processes of trialling and piloting to the model of active interviewing discussed above (Holstein and Gubrium, 1997). These processes seem geared to getting the right set of questions to yield the right information from an interviewee. That said, Gillham (2000) is right to be concerned about the importance of crafting good questions although this might come from a reading of the literature as much as it might derive from trialling.

Scholarly Preparation

You need to prepare for an interview by reading as much as you can about the issue so that when you hear or offer provisional ideas, connections and promising conceptual hooks, your sense-making is from a scholarly basis. This reading will also help you to formulate the direction of your questions for the interview schedule. Whether you then trial these questions or explore them in your first interview very much depends on your confidence, experience and on practical questions.

Formulating the Interview Schedule

The semi-structured interview schedule is always a working document. In the course of interviews, you might add to or amend questions as interviewees alert you to the need to do so. The schedule is constructed to facilitate a good conversation and inquiry. If you ask too many questions or take a checklist approach to your prompts, the event will become too researcher-centered and too structured. Aim for Burgess' (1988) image of interviews as "a conversation with a purpose" and my image of a "third space" for inquiry.

Conducting the Interview

Shank (2002) suggests that if we are good at conversations (e.g. listening, encouraging, emphatic, prompting, etc.), we are likely to be good at interviews. Similarly, Rubin and Rubin (2005: 81) suggest that you think about how your own style can be extended or adapted:

> If you are too intense, lighten up by introducing a little chat. Or, conversely, if you are too gregarious and have too much polite banter at the beginning of an interview, work out a shorter introduction and stick to it. If you are anxious and find yourself asking only the questions that you have written out in advance regardless of what the interviewees are saying, accommodate by writing out only a few questions and force yourself to wing the rest by listening and responding to what you hear.

Gillham (2000: 35) is more cautious about harnessing conversational skills:

> Everyday "conversation" is often a kind of jostling, with the nominal listener more or less impatiently waiting for his or her turn . . . becoming a listener rather than a talker is the biggest single problem in interviewing training.

Perhaps it is best to think of the interview as an interactive performance that needs some preparatory scripting and stage management. The interviewer has to acquire a repertoire of questions (prepared and impromptu) for encouraging a good, exploratory conversation. Technique needs to be accompanied by a scholarly understanding of the field so that the formal questions and the ad libbing are not in a theory-free zone.

Main Questions, Probes and Follow-Up Questions

Rubin and Rubin (2005) offer a structure for the interview questions that breaks down into three categories, namely:

- main questions;
- probes; and
- follow-up questions.

In my experience, probes and follow-up questions can overlap considerably but this does not diminish the value of thinking about these three kinds of questions for the planning and the conduct of a good interview. Rubin and Rubin's (2005) "responsive model" provides very helpful ways of opening up the interview talk and of ensuring developmental space throughout the conduct of the interview.

Main Questions

A semi-structured interview has between five to eight main questions. These questions will derive from the themes identified in the trialling stage. For instance, themes for an exploration of plagiarism might include:

1. the study traditions of interviewees;
2. assessment experiences (getting grades and feedback on assignments);
3. attainment experiences;
4. perceptions of collaboration; and
5. perceptions of cheating.

Once you have identified themes like these, you will formulate main questions against them. In the example above, the first could be:

"I am interested in the different ways in which schools prepare us for university. Can you take me through your final years of secondary school?"

Grand or Mini Tours

The above is a "tour" question because it gets the student to guide you through a specified experience (e.g. the last time you lectured? a typical seminar? Yesterday? Your first week on campus? How you set up the classroom? Your first day of vacation?). Tour questions are common openers because they tend to be unthreatening and a fruitful way of getting interviewees to talk about their experiences.

Highs, Lows and Iconic Moments

A further fruitful main question asks the interviewee to identify experiences or events which stand out in some way. Like Rubin and Rubin (2005: 161), I find this kind of question often yields very rich and unexpected data. Similarly, eliciting "Iconic moments" (Rubin and Rubin, 2005: 132) can produce vivid accounts. To illustrate, against the second and third main themes above, you might ask:

- Have you ever been given an assignment that really stood out as stressful or difficult to deliver? Can you take me through how you tackled this?
- Have you received a grade and/or feedback which became a turning point for how you saw your future achievement?

Hypothetical Questions

Further suggestions from Rubin and Rubin (2005: 161) include posing hypothetical questions. These can take the form of an imagined future:

- If you were in charge of designing the graded assignments for this module, what would they look like?

Or to explore a sensitive area such as theme 5 above:

- How do you think a university can discourage students from copying other people's work?

Compare or Contrast

Asking interviewees to compare and contrast two sets of experiences prompts them to think about what is distinctive about each:

– Can you tell me about your experiences of getting a high and a low mark respectively?

Task Question

This involves asking the interviewee to do something as the basis for discussion, e.g.:

– Can you visually depict what it was like to be investigated for academic misconduct?
– Can you read this piece so that we can talk about it?
– Can you look at these photos and tell me what they mean to you?

Probes

The purpose of probes is to "help you manage the conversation by regulating the length of answers and degree of detail, clarifying unclear sentences or phrases, filling in missing steps, and keeping the conversation on topic" (Rubin and Rubin, 2005: 164). You need to be aware of a range of possible probes and to develop the skill of knowing which one to use as the interview progresses. Here are some examples —for a fuller discussion of probes see Rubin and Rubin (2005):

• That's very interesting. Can you tell me more?
• That must have been difficult. Can you tell me more?
• Do you have an example of when that happened?
• Did that happen often? How many times?
• Could you clarify? Perhaps with an example?
• What does "x" mean?
• Then what happened?
• How do you see that as related to the topic?
• We can come back to that later if you prefer

Other probes can be non-verbal such as nods, "uh ha's" and encouraging silences to signal that you want to hear more.

Follow-Up Questions

The success of the interview centers on the interviewer's ability to follow hunches, hints and openings through good follow-up questions. As indicated, there is an overlap with probes and follow-up questions because the former sometimes prompts the interviewee into talking more deeply about a topic which is the explicit purpose of follow-up questions. Some follow-up questions might take place in a second interview. Others are formulated from the identified gaps in the transcripts of those already interviewed.

Getting More Detail

You cannot build up a rich picture of your topic if you accept broad sweep responses to your questions. Some of your follow-up questions will be designed to get at the detail. In these cases, the follow-up question will ask for more information, some detail, an illustration, an explanation, an exception to the rule, etc. (Rubin and Rubin, 2005).

New Ideas

Where an interviewee has said something that other interviewees have not raised, you will want to check its value for your research with questions such as:

- Is that a widely held view?
- Do other students feel the same?
- You are the first to raise this, can you think why?
- Do you see your experience as different to the others?

Checking Understanding

Gillham (2000) advises that the interviewer develop the therapist skill of reflecting back to the interviewee what they understand has been said, with intermittent paraphrasing such as "so you feel that your workload prevents you from talking to individual students?" Take care that this strategy does not invite interviewees to agree to settled positions in cases where they may be ambivalent.

Stories

When interviewees begin to tell stories in response to a question, it is always worth asking follow-up questions to secure a fuller narrative because it is likely to offer rich data. Your questions are likely to be "chronological" or "stage" ones to progress accounts of the sequence of events:

- What happened after that?
- And after secondary school?
- So what happened when you left university?

Again, see Chapter 6 on narrative inquiry for a further discussion about eliciting stories and critical events.

Exploring Positionality

If you want to keep a strong "active interview" line of inquiry (Holstein and Gubrium, 1997), you might explore with the interviewee a possible connection between positionality and the account provided, e.g.:

- You started off by saying that you had never been to England before you arrived at university in London. Do you think any of this would have happened to you if you had been an English student?
- How would your parents make sense of this?
- Do you think your teachers would take the same view as yourself?
- How do you think they would see it?

Ambivalence, Contradictions

It is worth exploring any hesitancies, tensions or contradictions in interviewee responses to discover what lies beneath them. If a student says she is not too sure that assignment marking is fair, left to itself, this is tantalizing but unhelpful data. You will need to follow up with a probe such as:

- Can you say a bit more about that?

If she continues to be hesitant, you might want to go for a mini-tour follow-up question:

- Can you take me through what happens when you get a disappointing mark?

Theorizing With the Interviewee

As the interview proceeds, you may want to formulate your own provisional explanation of an event or conceptual understanding of the topic and share it with the interviewee in the spirit of an "active interview":

- From what you are saying, it seems that your first semester at university was a transitional state where you were still strongly connected to home but also developing a new home with other students. So you seemed to have been pulled in two directions. Does that sound right?

If the answer is tentative, you might follow it with something like:

- Perhaps there is another way of looking at this experience? Do you have any thoughts on that?

Drawing on Scholarship

If the interviewee says something that resonates with what you have read in the literature, you might want to bounce this off them:

- Some educationalists say that universities give students assignments to test what they have learnt rather than to help them to learn. So it's a testing game rather than a learning mechanism. Is that what you are saying?

You have to be watchful about tiring the interviewee with difficult questions, particularly towards the end of the time; if you sense that you are doing this, you can always add:

- You might want to think about that and get back to me later.

Closing the Interview

As the interview time runs out, plan to end punctually. If you think that the interviewee has more to say, you can arrange to meet again or to have a follow-up telephone conversation or email exchange. Ask the interviewee if he or she has any questions. I always ask a final sweep question such as:

– Is there anything that our discussion hasn't covered that you would like to add about this issue or about the research project?

Finally, you thank the interviewee and repeat that he or she will get a transcript; specify a deadline by which you would like comments back, failing which, you will assume consent to quote from it. If you are looking for more interviewees, this is your chance to ask if he or she knows of any likely volunteers (this is called snowballing).

Research Diary

Once the interview is over, you write your own field notes about the event to add to your data. This will include reflections on the questions and answers, gaps you think you didn't fill, any feelings you had about the setting and interviewee and so forth.

Telephone Interviews

Telephone interviews can also be used for semi-structured interviews. My own initial scepticism that a telephone interview is a poor compromise was challenged by the experience of rich discussions with academics across six universities. These are the important principles I learnt from the process:

• Send the interviewee a brief schedule of questions.
• Fix the telephone appointment to allow for some quiet time before (to go over schedule) and after the discussion (to write field notes).
• Put a "do not disturb" sign on your door.
• Organize and test the technology for telephone recording (very cheap and very effective).

- Do not go over forty-five minutes (telephone interviews are very tiring).

Some interviewees prefer telephone interviews because they feel more comfortable disclosing through this medium. Internet interviewing is also a growing field.

Transcription and Analysis

Interviews do not have to be recorded though note taking is tiring and may interfere with the conversation flow. Most transcripts are verbatim and include pauses, repetitions and idiomatic expressions (such as "innit" or "you know"). Walford (2001) argues against complete transcription, preferring instead to listen to the recordings and to take notes from them. This enables Walford to have a stronger sense of the interview event and to avoid treating people's testimonies as their last and definitive word on the topic.

In displaying some of the transcript in your report, use the quotes convincingly: they must not look laundered or cherry picked. Incorporate your comments to show the developmental dynamic of the event.

Holstein and Gubrium (1997: 127) write that the analytical process for active interviewing involves showing the dynamic interrelatedness of the *whats* and the *hows*. This is achieved by combining attention to the developmental dynamic of the interview (using some of the moves I have suggested in Chapters 3 and 6) with attention to what is being said about the topic under investigation. Holstein and Gubrium (1997: 127) explain:

> The goal is to show how interview responses are produced in the interaction between interviewer and respondent, without losing sight of the meanings produced or the circumstances that condition the meaning-making process. The analytic objective is not merely to describe the situated production of talk, but to show how what is being said relates to the experiences and lives being studied.

The last sentence of this quote is key. You are aiming to say something about *how* the meanings were produced and something about *what* these meanings might tell us about the topic in hand.

See Chapter 3 where I discuss the "how" of analysis through attention to "contrastive rhetoric," "oppositional talk," "stake inoculation," "membership categorization" and figures of speech.

Conclusion

I hope I have clarified that before you think about what you want to ask interviewees, you need to come to a decision about what you think the interview can accomplish. Will you be "prospecting" information, negotiating meanings or something in between? Your decision will determine a number of the moves you will want to make both in the interview and for the analysis of the data you gather.

Further Reading

Alldred. P. and Gillies, V. (2002) Eliciting research accounts: Re/producing modern subjects. In M. Mauthner (Ed.), *Feminist ethics in qualitative research* (p. 146). London: Sage.

Burgess, R.G. (1988) Conversations with a purpose: The ethnographic interview in educational research. In R.G. Burgess (Ed.), *Studies in qualitative methodology: A research annual* (Vol. 1, pp. 137–155). London: JAI Press.

Fontana, A. and Frey, J.H. (2000) The interview: From structured questions to negotiated text. In N.K. Denzin and Y.S. Lincoln (Eds.), *Handbook of qualitative research*. London: Sage.

Gillham, B. (2000) *The research interview*. London: Continuum.

Holstein, J.A. and Gubrium, J.F. (1995) *The active interview*. Thousand Oaks, CA, and London: Sage.

Holstein, J.A. and Gubrium, J.F. (1997) Active interviewing. In D. Silverman (Ed.), *Qualitative research: Theory, method and practice* (pp. 113–129). London: Sage.

Holstein, J.A. and Gubrium, J.F. (Eds.) (2003) *Inside interviewing: New lenses, new concerns*. Thousand Oaks, CA, and London: Sage.

Rubin, H.J. and Rubin, I.S. (2005) *Qualitative interviewing: The art of hearing data*. Thousand Oaks, CA, and London: Sage.

Schostack, J.F. (2006) *Interviewing and representation in qualitative research projects*. Buckingham, UK: Open University Press.

Shank, G.D. (2002). *Qualitative research: A personal skills approach*. Columbus, OH: Merrill Prentice Hall.

Spradley, J.P. (1979) *The ethnographic interview*. Fort Worth, TX, and London: Harcourt Brace Jovanovich.

Walford, G. (2001) *Doing qualitative educational research: A personal guide to the research process*. London: Continuum.

6

NARRATIVE INQUIRY

Appeal

Narrative inquiry is particularly useful if you want to know something about how people make sense of their lives through the selective stories they tell about noteworthy episodes. For instance, all learners arrive at their place of study with stories about how they got there. These might be stories about triumph over adversity, parental pressure, ambivalence about options and so forth. Gathering and exploring these stories will allow the researcher to gain insights into the complex ways in which learners come to make decisions and to act on the basis of them.

Narrative inquiry offers a good way of probing into how learners and academics perceive opportunities, conflicts, compromises, ambitions, values and professional or study activities; it allows the researcher to grasp the diversity and complexity of personal choices over time and across a group; it also allows for the identification of commonly shared positions and meanings (Reissman, 1993; Savin-Baden, 2004; Webster and Mertover, 2007). Narrative inquiry for higher education research can be used to:

- explore development trajectories among learners or teachers;
- explore transitions among learners (e.g. from school to college);
- gain insights from learners about particular events with respect to issues such as plagiarism, perceived discrimination and marginality; and
- generate understandings about particular learner experiences of pedagogic practices (tutorials, problem-based learning, etc.)

93

Purpose

It's the Way We Tell Them

Just as authors narrate journeys for their characters which are punctuated by critical events, plots and subplots, so do we in making sense of our lives (Reissman, 1993; Clandinin and Connelly, 2000). Similarly, just as authors take bits of truth and fashion them into a story, so do we in narrating our past. That is not to suggest that we all consciously lie but in selecting what to tell and how to tell it, we inevitably change the events. While most of us try to adhere to what we think really happened, we can never entirely replicate this in narrative form. For this reason, narrative inquiry is not simply interested in the truthfulness of narratives, rather it is concerned with how we make sense of and structure our experiences from the stories we tell.

To illustrate, if we are asked to provide stories of our formative schooling experiences, we are likely to package them into a narrative that inevitably excludes some parts of our experiences, perhaps embellishes other parts and perhaps even falsifies others (not necessarily consciously). For instance, we might provide a narrative plot that centers on parental disappointment about our less spectacular achievements at school. We might offer a critical episode when this happened. Even if this is half the story because our parents were also very encouraging, the fact that we have singled out this slice of experience hints at the overriding potency of perceived negative parental feedback. As well as the question of selectivity of this kind, narrative inquirers must be alert to the language the storyteller chooses (of which more under Method).

In many respects, the subtext of the narrative is of more importance to the researcher than the explicit text (Savin-Baden, 2004). For this reason, as we shall see, the analysis of narratives explores both *what* people say and *how* they say it. In sum, the purpose of narrative inquiry is to interpret both the form and the content of the stories gathered in order to generate understandings of how personal histories influence the narrators' values, decisions and actions.

Theoretical Concerns

Storied Selves

A key premise of narrative inquiry is that we are all storied individuals and, as such, we carry around a repertoire of explanatory and justificatory stories about our experiences. As we go through life, we may edit, adapt, extend, reject or repress some of these stories but we will never be without storied versions of who we are. The truth of this was brought home to me when I once interviewed single parents who had returned to study. All of the women offered vivid and elaborate stories about the painful processes by which they prepared to separate from their partners. I had not asked the women to provide stories of these processes but they all found space in the interview to tell them. It seemed that these were stories told and retold both to themselves and to others in order to organize and articulate how difficult experiences were endured and overcome.

Sometimes we keep stories to ourselves or to those most close to us. Sometimes we might tell a personal story to exemplify a deeply held value or a moral point (Reissman, 2001). Someone might narrate how a member of the family went through a painful divorce in order to demonstrate the strength of their commitment to marriage. Just think how often people offer opinions through the example of something that "happened to them." The narrative inquirer, then, prompts our natural impulses to enter into story telling.

True Stories?

Although there is no such thing as a true story for narrative inquirers, they do expect that the stories narrated will have *some* relationship with what has happened to the narrator. While the analytical process is concerned with what the narrator has selected and the manner in which he or she has told the story, their accounts are not treated as pure fictions. The researcher is required to demonstrate the trustworthiness of his or her data gathering and analysis and the plausibility of the stories gathered (Webster and Mertova, 2007). This is secured by the writing of a "persuasive" account (Reissman, 2001) that presents enough of the data to allow the reader to be convinced of the interpretation made. See Savin-Baden's (2000) collection of "stories untold" in relation to problem-based learning as a good example of such an account. There is also the question of researcher reflexivity, and below I will present Savin-Baden's own model for analyzing stories to maximize a reflexive dimension.

Webster and Mertova (2007: 101) draw on Lincoln and Guba's (1985) notion of "transferability" as a form of generalization. This rests on the view that the narrative inquiry report must provide a rich depiction and analysis such that a reader can explore comparisons in another setting. This view is very similar to those put forward, largely from Stake (1995), in the chapter on case study research. Similarly, Webster and Mertova (2007) talk about the comparison of critical episodes in stories with "like events" across narratives. Indeed, their approach to narrative inquiry is very mindful of a quantitative dimension to the gathering and analysis of data.

Back to the Future

For narrative inquirers there is an intimate link between how we look back and how we look forward. The stories we construct about our past shape how we see ourselves and what we think it is possible to do with our lives. We are all historians of ourselves and the task of the narrative inquirer is to facilitate the telling of our history to uncover critical events or turning points (Webster and Mertova, 2007; Reissman, 1993). To complicate matters, we are likely to carry more than one historical version of ourselves and which one we select to tell will have much to do with who is listening and where and why they are listening. After Bakhtin, this contextual dimension has come to be understood as a problem of "addressivity" (Cheyne and Tarulli, 1999).

Addressivity

What we say is determined by what has already been said (Bakhtin calls this the history of utterances), the context in which we say it and our audience. This raises an interpretive problem of addressivity for narrative inquirers. For instance, I might tell a story in embellished form to amuse dinner party companions; I might tell a story to a new employer from the defended position of someone who has been sacked from my former post; I might tell a researcher an ego-centered story that places me in a good light; I might sense that the researcher wants me to talk about a wounded past and thus give him or her a victimist account. This variety of possible histories does not make the task of the researcher hopeless; it merely heightens the need for reflexivity about the context of the story telling and the possible power relations or agenda between researcher and narrator.

Story Spaces

Even if the researcher keeps her questions to a minimum to allow the free flow of the story, he or she must bear in mind that story telling is a cultural and emotional event—indeed one of our oldest— in which both teller and listener are drawn into a story "space." Think of a parent telling a child a bedtime story. There are pre-story rituals of choosing the story, snuggling into bed, preparing for the story's historical reach ("once upon a time"), dramatizing the plot (the narrator changes tone and pace to signal high drama) and expecting the relief of a denoument ("and they all lived happily ever after"). The researcher needs to be reflexive about what went into the construction of the story space and what might be influencing the direction of the story.

Telling the Whole Story

Because the story is the unit of analysis, narrative inquirers try to resist fragmenting the story according to dominant conventions of qualitative data analysis (Savin-Baden, 2004). To illustrate the problem with such conventions for narrative inquiry, you could, for instance, take Cinderella's narrative of her story and generate a set of descriptive codes such as: poverty, sibling rivalry, slippers, beauty, domestic exploitation, step parenting, second marriages, princely rescue, pumpkins and coaches. You might then move to a more analytical stage and come up with, say, attributes of an unhappy second marriage and bullying. But in fragmenting Cinderella's narrative in this way, have you disturbed its rhetorical drift about the triumph of good and virtue over evil and dissoluteness? And will codification lend itself to the retrieval of a victimist subtext to Cinderella's story? Further, a story has a beginning, a middle and an end (though postmodernists might quibble with the linearity here) and chunking data can undermine the historical coherence of the narrative. Cinderella was happy until her mother died. She became exposed to stepfamily abuse when her father went away on business. The sequencing of events of this sort is a crucial part of the structure of stories. Just a consideration of how often narrators use the term "and then" in their storytelling will demonstrate the importance of sequencing. A practical guide to preserving the coherence of the story is elaborated below.

Method

Narrative Inquiry Research Questions

Like any other qualitative research question, narrative inquirers have in mind a specific puzzle, a conjecture or an evaluation question such as the following:

- Have faculty developers increased their effectiveness over the years?
- Why are geography students achieving lower marks than other students?
- Given their dual commitments, how do mature students manage to do so well?
- What worth can be assigned to the curriculum changes implemented in this course?

All of these questions and puzzles are amenable to narrative inquiry because this research approach is likely to identify illuminating and telling episodes. Of course, you need to be confident that people will have a history in the area you want to probe. Clearly, it would not work to ask undergraduates in their first week of study to provide stories about their learning trajectory at university.

Another kind of narrative inquiry question might concern what sociologists call the "moral career" or the "identity formation" of individuals. To illustrate, a university student is formally anyone who registers for a degree but this formality only inaugurates his or her passage to studenthood because processes of enculteration have to take place for him or her to *feel* like a student. Collecting their stories about their undergraduate years will enable us to surface important formative experiences for the formation of student identities.

You can also use narrative inquiry to explore how students experience pedagogic methods as in Savin-Baden's (2000) research into students who have studied through problem-based learning.

How Many Stories Do You Need?

You might want to try for a range of storytellers within a group but like all qualitative research, the goal is not to secure a representative sample but to generate understandings from going deep rather than wide. The research report will offer a brief biography of the storyteller (e.g. age, sex, family background) though, of course, as with any research, glib associations with this and the content of the story told are to be resisted (which does not mean that no associations can be made).

In judging how many stories to collect, pragmatically, you have to judge what time you have available for arranging access, gathering, analyzing and writing up your analysis. Theoretically, you can treat each narrative as a "case" and following the rule of case study research about the range from single to comparative case study although a single narrative is not entirely equivalent to a single case. In my view, you want at least five narratives from a particular group in order to say something plausible and compelling about it. If you have the time, try for ten narratives (or even more) but not if it means sacrificing depth for breadth.

Interviewing for Stories

There is no set formula for getting narrators to tell their stories but since you want the storyteller to do most of the talking and you want the talking to be centered on the past, you need to give some thought as to how you are going to prompt this. Some of the suggestions I make below will not neatly apply to the particular context in which you are eliciting stories. For instance, if you have a shared experience with the narrator, you might be able to tell some of your story (without hogging the attention) as a warm up. Some of the plots to the stories might be traumatic or stressful while others might be less emotionally charged—questions will need to respect this variation. Each story space will be unique and you have to follow your nose about what to ask and when to stay silent. The only general rule across all contexts is to listen more than to speak.

The narrator might be expecting a more conventional interview so you do need to clarify that you are collecting accounts which range from small incidents to quite significant events. Some people find the term "story" confusing because they associate it with fiction and you might need to indicate what you mean by this.

Unless consent has been refused, you will want to tape (or video) the interview. If you have never collected stories before, craft your skills at narrative inquiry by trying them out. In this respect, you are not so much testing the "right" questions (a convention of trialling) as learning how to encourage story telling, bearing in mind the following.

Creating a story space   Does the interview environment facilitate storytelling? Is it comfortable and welcoming? Will you be disturbed? Is there sufficient time for the narrator to ease into story-telling? (you can't rush the interview). Present your interests and ethical framework (e.g. sharing your analysis). Do what you can to reduce the researcher/researched distance—perhaps say something about why you have a personal investment in this area. Test the tape recorder (or video) at the right moment. Say how long you expect the interview to take (two hours maximum) and how much you value the time your narrator has set aside to tell his or her story.

Which history?   Decide the time span over which you want accounts. If you are interviewing postgraduate students, do you want them to provide stories over their entire supervised experience or do you want to focus on a particular historical phase? Can you construct a storyboard to facilitate the narration (Reissman, 1993)? This would identify moments along a trajectory against which you ask questions:

a) when you first thought about doing postgraduate work; b) submitting a proposal; c) early supervision; and so forth.

Visuals or objects?   Some narrative inquirers use strategies associated with reminiscent studies where the storyteller is shown pictures, photos or objects that stimulate their thinking. These may be from the storyteller's own collection or they may be selected for them. You might, for instance, show a group of international undergraduates a series of photos which depict their induction activities and ask them to reflect back on these. You might also ask them to bring something to the interview which symbolises their feelings about being away from home and family.

Warm up questions   Given that story telling often requires high levels of disclosure, some initial, unthreatening, framing questions can begin the process. Ideally, these questions can be formulated to get the teller to think back in time. They resemble the first questions of a story board approach. Let us assume your aim is to capture stories from students who have been accused of plagiarism, you might start with the following "first" experience type questions (within the timeframe you have set):

- Can you take me through your first semester?
- Can you say a bit about how you met your first deadline? (or not)

Plot Questions

When you sense that the teller is comfortably delving into his or her past, you can begin to center on "plots" to their stories (though some narrators will have moved spontaneously to them) with questions such as:

- Can you think of a time/moment/episode that stands out? (negative/positive and personal/academic)
- Can you reflect back on any high points over this period? And low points?
- Can you comment on any event that changed how you saw things? Or changed how you acted? Or felt?

- Can you think of something that happened which seemed trivial at the time but made you think of it as more conse-quential later?
- Can you think of anything that happened that started to nag in the back of your mind?
- Do you have a sense of unfinished business about anything that has happened?

Give the narrator plenty of time to think about one of these questions and tolerate a reasonable amount of silence while they are doing so.

The last four questions above are aimed at exploring the significance of the apparently trivial. For instance, a student might brush off a feeling of disappointment about a mark that comes to haunt him or her later. She/he cannot define a specific episode because her disquiet emerged over time.

Some of the questions will help you to gather critical incidents (Webster and Mertova, 2007). You will want to check that these incidents cover:

- When and where the episode took place.
- Who was involved.
- The detail of what happened.
- How they felt about what happened.
- Why they think it happened.
- What made it particularly noteworthy.
- What they did and felt after the episode.
- Whether similar episodes have occurred. Whether these are common to other individuals or groups.
- The influence of the episode on their present values and/or actions.
- Whether they have a shared interpretation of the events with others involved in it or in similar events.

Wait till the teller has provided a break in their narrative before asking for any clarification you need. The challenge will be to cleave to the storyteller, to appreciate that he or she has probably answered most of the above questions (and some you did not think of) and to keep well away from any connotation that you are interrogating. Think armchair, not witness box. At the same time, it may be that some of the questions you ask will be with an eye on the storyteller's positionality.

Bear in mind that if we are emotionally invested in a story, we usually tell it from a defended position. To give a banal example, if you have bought an expensive car and someone asks you whether it was worth the financial sacrifice, you might well reply "yes" because you need to defend the expense to yourself and to others. You might keep your buyer's remorse to yourself. Possible questions to get the teller to rethink a defended position, might be to have them consider rival explanations or alternative steps they could have taken:

- If you could retrace your steps to the beginning of this story and replay your own actions, how might they be different?
- Were there alternative steps you might have taken?

Remember, you are not trying to extract the truth but you do want the narratives to be thoughtful, which includes exploring tensions and contradictions in the first telling of them (Reissman, 1993). Similarly, check that you are not inviting the stories you want in order to fit them into your own preconceptions or experiences.

Reflexivity in narrative inquiry is very much a two way thing in that the researcher is being reflexive about his or her positionality and the suggestiveness of her questions while supporting the narrator in being similarly reflexive about her stories.

Winding Down

Once you sense that the narrator has said as much as she or he wants, that you are close to an agreed finishing time and/or that it would be tiresome/tiring to her to deal with more questions, signal that you are mopping up with questions like:

- Your story/ies are giving me a fascinating insight into this issue. Do you think there is anything else I should have asked you? Or that you want to add?
- If anything occurs to you afterwards, here is my email. I'd welcome further comments.

– Are there any other people you know who might be willing
to talk to me?

Finally, if you have decided to share your transcript and/or
interpretation with the narrator, you will want to make arrangements
for that to happen.

Commissioning Stories

Most narrative inquiry methods involve an interview, although there
are other ways of gathering stories. For instance, Bill Dunn (2002),
asked his first year engineering students to write stories about their
initial weeks at the university. His aim was to discover more about
how they experienced their first semester. His worry was their apparent
lack of motivation. For a given period each week, towards the end
of a regular lecture, he asked the students to write stories to a given
focus. Alternatively, stories can be collected via email.

Analysis, Interpretation and Reflexivity

There are diverse strategies for analysing narratives; some use a socio-
linquistic approach (for a helpful guide to this see Reissman, 2001
and Grbich, 2007). Below I will present a method generated by my
former colleague Maggi Savin-Baden who has generously offered the
following interactionist–interpretivist approach. This approach tries
to overcome the "proceduralness" (Savin-Baden 2004: 3) of data
analysis by achieving a "shift from lists and codes to understanding
the subtext of data" (2004: 2) and by embedding reflexivity into the
interpretive process. Here are the stages:

1. Once you have gathered your narratives, set them to one side
   and write a short biography of each narrator on the basis of
   what you remember from the telling of the story. This need
   not be more than one or two sheets of A4.

2. Now reread the transcripts and listen again to the audiotapes
   in the light of the biography and explore the following
   questions:

a) Are there any particular discrepancies between your account and the narrator's?

b) What do you seem to have privileged?

c) What do you seem to have omitted?

d) Which quotes/concepts from the interview support or challenge the biography you have written

An exploration of these questions will increase your reflexivity about the interpretations you are making as well as increase your intimacy with the data. It will also give you a provisional idea of the thematic commonalities across the stories—to which you will return later.

3. These first steps of analysis help you to think about what you bring to the interpretation through a focus largely on *what* the narrator said. Next you will want to look at *how* the story was told, paying attention to the way in which language is used. To illustrate briefly, here is an extract from one of the stories Bill Dunn (2002) collected during induction week for first year engineering students at a British university. The students were asked to write a story about living with other students:

> Both my house mates are very messy. One of them thinks that plates wash themselves and that the bathroom never needs cleaning, ever. "There are magic fairies that come down every night and do it for you," he thinks. The other messy house mate is a little better, he realizes that the house actually does need cleaning occasionally but doesn't do much about it. As soon as he brings any of his mates over, an amazing coincidence occurs. At the very time they come over a bomb goes off in the living room and little elves come into the kitchen and steal all my bread and butter, amazing really.

Here are two things that might be said about how this student tells his story:

a) His virtues as a tidy man are expressed by contrasting himself with his "messy" housemates. The language move here is to use "oppositional talk" to guard against the "noxious identity" of messy housemate (Savin-Baden 2004: 8).

b) This student's use of "magic fairies," "little elves," "bomb goes off," "amazing really," draw on the kind of heavy sarcasm parents often use when admonishing children for being messy; by drawing on a parental script of this kind has this student assigned to himself a parental identity?

If you just look at the content of the extract, it tells you something about messy students. Exploring *how* the story is told, its subtext, extends the interpretation, allowing the researcher to consider, for instance, the extent to which the storyteller might be defending a particular stance.

In this phase, then, you have explored the narrative structure, figures of speech and rhetorical moves the storyteller may be using. You are particularly looking at how the storytellers are positioning themselves. Have they defined themselves in the story (e.g. "as an expert," "as a biologist")? Have they used the passive to distance themselves (e.g. mathematicians tend to see themselves at the top of the academic pile)? Have they dissociated themselves from a noxious identity in order to say something from within that identity (e.g. I am not a racist but . . .)?

You are also looking at what figures of speech narrators are using. Savin-Baden (2004: 12), for instance, writes about how her typification of a research intensive university as a "sausage factory" skewed her vision of this institution. Acquiring the skill of recognizing subtextual clues in people's stories is a matter of practice and attention. Savin-Baden entreats the researcher to share interpretations with the storyteller so that the sense-making of the story is a mutual enterprise.

For further ideas about analysing the "how" of the narrative, see the chapter on data analysis where I discuss metaphor, "contrastive rhetoric," "stake inoculation" and "membership categorization" as things to watch out for in your narrative.

4. In the light of your interpretation, the next step (which overlaps with 3) is to look at each story and ask:
   a) what holds this story together?
   b) how does what was said link to how it was told?

    c) where are the conundrums and things that seem to be at odds with one another?

If you can go back to the narrators to explore this, you will be strengthening both your and the narrators' capacity to articulate what might be going on.

5. Now you rewrite the biography in a longer version which takes into account your analysis and interpretation since the first biography and add in quotations from the interview.

6. Now you look across all the interviews, asking "what holds these stories together?" Try to identify four or five overarching themes that are common across the stories. Be mindful too of what is *not* said and of anything that does not fit into overarching themes.

Writing Up

By following the above process, you are beginning to write up as you analyze. Indeed writing is part of the analytical process. You will need to come up with a structure that preserves the integrity of the stories but also gives you space to present your interpretations. Although you may want to say something about your own positionality with respect to the research, take care that you do not end up writing about yourself more than you explore the lives and meaning-making of the narrators.

Conclusion

If your research focus lends itself to the collection of stories, live or online, commissioned or gathered through interview, this is a good research method to choose. Basically, if you want to know more about the lives and experiences of particular groups of learners or academics, narrative inquiry is a good choice. Whatever method of gathering and analysing you choose, you will need a fair amount of time to do this kind of research.

Further Reading

Reissman, Catherine Kohler (1993) *Narrative analysis, qualitative research methods*, Series 30. Newbury Park, CA: Sage.

Reissman, Catherine Kohler (2001) Analysis of personal narratives. In J.A. Holstein and J.F. Gubrium (Eds.), *Inside interviewing: New lenses, new concerns* (pp. 331–346). Thousand Oaks, CA, and London: Sage.

Savin-Baden, M. (2000) *Problem-based learning in higher education: Untold stories*. Buckingham, UK: SRHE/Open University Press.

Savin-Baden, M. (2004) Achieving reflexivity: Moving researchers from analysis to interpretation in collaborative inquiry. *Journal of Social Work Practice*, 18(3), November, 1–14.

Webster, L. and Mertova, P. (2007) *Using narrative inquiry as a research method: An introduction to using critical event narrative analysis in research on learning and teaching*. London: Routledge.

7

ETHNOGRAPHIC APPROACHES

The Appeal

Ethnography simply means the description of groups. At first sight, it might seem that anyone can do ethnography but doing it well requires familiarity with a theoretical field, a set of research skills and perhaps, above all, to use Eisner's (1991) term, an "enlightened eye." Ethnographic approaches appeal to those who are confident that they have the time to stay in a research setting (the field) for at least a couple of weeks (it might be where they work anyway) for sustained observation and informal interviewing.

The Purpose

Ethnographic research derives from anthropology where it largely concerns long periods (a year or so) of fieldwork with specific communities or groups. Key methods are observation and informal interviewing of "informants" for the "purpose of learning from their ways of doing things and viewing reality" (Agar, 1980: 6). Here is a summary from Delamont (2002: 8):

> The central method of ethnography is observation, with the observer immersing himself/herself in the "new culture." Ethnographies involve the presence of an observer (or observers) for prolonged periods in a single or a small number of settings. During that time the researcher not only observes, but also talks with participants.

Spradley (1979: 3) stresses that ethnography is not so much about studying people as learning *from* them and this is why ethnographers use the term "informant" rather than "subject" or "respondent." The informant is teaching the ethnographer. More broadly, Agar (1986) describes ethnographers as mediating between two cultures, that of their own and that of the group studied. The differences between the two constitute a "breakdown" (Agar 1986: 20) which the ethnographer must seek to understand and explain.

Ethnographic research is often carried out in places that are far flung for the researcher or among marginal groups such as drug-takers and mental health patients but it can equally apply to mainstream groups (e.g. academics and student groups) and to close to home settings. Likely fields for ethnographic approaches in higher education might be:

a) academic cultures from the viewpoint of a discipline, faculty or university;
b) particular student groups in specific settings (fieldwork, laboratories, lectures, libraries, residences, programs, campus);
c) academic management groups; and
d) online communities and learning environments.

Wherever the research setting, it is never purposefully manipulated as in, for instance, experimental research. Ethnography addresses a worry that research into how people behave in experiments might only tell us how people behave in experiments. Ethnography takes a holistic view of the research setting (the field), arguing that the premature selection and isolation of particular aspects of it will skew our understandings of complex, human contexts.

Ethnographers, then, research the world as it presents itself to them (this is called naturalist inquiry). Their aim is to explore the meanings underlying human behavior. Typically they gather data through a combination of conversation, observation, participation and interviewing; they might also use videos, cameras or drawings as well as informant diaries. They record their observations, thoughts, scholarly references and theoretical elaborations in field notes and in a research diary. Data gathering, data analysis and writing-up proceed alongside each other in a process that involves continual focussing, reflexivity, scholarly engagement and theory building.

Netography

There is an emerging field of netography. Here the ethnographer either participates in virtual groups and communities (which can include second life) or "lurks" as an observer. Some researchers are also combining online with face to face encounters with informants. This field involves significant adaptations from conventional ethnography to accommodate the fact that the field is not a bounded "place," that people meet asynchronously and that there are distinctive ethical issues associated with netography. See Hine (2005) for further guidance and explorations into this area.

Visual Ethnography

Another relatively new field is that of visual ethnography. While there is a long history of the use of video, photos and artefacts in anthropology, this new field seeks to privilege the visual as the site of explication rather than to see it as supplementing and thus in the service of data based on speech or text. See Pink (2001) and Stanczak (2007).

Autoethnography

This is not simply autobiography by another name. The point of autoethnography is to examine significant experiences from the standpoint of someone who has been through them. In discussing Ellis' (1997 in Grbich, 2004: 64) autoethnography of her experiences of caring for her dying husband, Grbich shows how this postmodern, literary move can unearth subtle layers of meaning which an outsider looking "in" may miss. In the wrong hands, of course, autoethnography might degenerate into the production of self-indulgent accounts.

Theoretical Concerns

Emic/Etic

A particular binary associated with ethnography is that of emic and etic. "Emic" research provides an insider view of what is going on in a setting and the "etic" a more detached and outsider view of a setting. Experimental design is etic and participant observation is emic. We should not overstate this distinction because an ethnographer is best positioned simultaneously as an insider *and* an outsider, as the section on reflexivity below indicates.

Macro/Micro

Yet another binary is that of macro and micro studies, with ethnographic approaches being associated with the latter, though, as Marcus (1998) has argued, even apparently cut off places are connected to dependent sites. There is no single locus of influence. For instance, a study of those infected or affected by HIV/AIDS in one part of rural Zambia will have to consider the urban areas to which some of the villagers migrate and then return with the disease; in turn this migration needs to be considered as part of a broader socio-economic reality for African countries. Ethnography is implicitly or explicitly multi-sited even though the focus may be in a single field.

Primary and Secondary Epistemic Seeing

Ethnographic studies aspire to generate a detailed, rich picture and theorization of the setting to support an understanding of both its specific contextual features as well as more generalizable ones. Eisner (1991: 68) describes these two related outcomes as the result of "primary and secondary epistemic seeing" respectively. This simply means that, for instance, an ethnography of academics in an English faculty in one university may well offer insights about faculties and/or about English teaching across other universities. This aspiration to produce local and generalizable understandings is shared by case study research which often uses ethnographic methods.

Decolonizing Ethnography

Ethnography has gone through a painful history of self-critique, particularly with regard to traditional anthropological studies which were thought to exoticize so-called "primitive" others from the standpoint of a colonizing gaze.

Contemporary ethnographic approaches sit in a variety of epistemological frameworks from positivist, naturalist, interpretivist, critical realist, critical ethnography, feminist to postmodern. To avoid being overwhelmed by the debates and conflicts that flow from these approaches, Hammersley and Atkinson (1983: 235) put in a plea against rigid paradigm adherence, arguing that what matters most is the adoption of a reflexive stance which is simultaneously about the research in hand and about its informing paradigm (Hammersley and Atkinson, 1983: 236). Further, Hammersley and Atkinson wisely caution against overdrawing distinctions between paradigms or of assuming a fundamentalist posture towards any one paradigm, as if it alone accessed the truth (even in the claim that there is no truth). In short, approach all "isms" with scepticism; take a provisional view about knowledge generation; and stay reflexive about what you are doing and the claims you are making.

Trustworthiness and Reflexivity

The advice of Hammersley and Atkinson is very sound, not least because nearly all contemporary ethnographers, whatever their differences, agree that they are the key research instrument and their interpretations are influenced by their own positioning. Thus nearly all ethnographers are big on researcher reflexivity and the quality of this reflexivity is acknowledged to be intimately tied into the trustworthiness of the account. Quite simply, you cannot do good ethnography without due regard to researcher reflexivity. Here is a good definition from Charlotte Aull Davies' (1999: 4) excellent book *Reflexive Ethnography*:

> Reflexivity, broadly defined, means a turning back on oneself, a process of self-reference. In the context of social research, reflexivity at its most obvious level refers to the ways in which the products of research are affected by the personnel and process of doing research. These effects are to be found in all phases of the research process from initial selection of topic to final reporting of results

In the Method section below, I offer some common areas of concern in relation to researcher reflexivity to support the researchers' writing of their diary and field notes.

Emergent Theory

Ethnography can be used to test and extend theories and/or to generate them. Either way, ethnography is inductive in that theorization emerges from the data. A broad research topic initiates the enquiry while more precise questions become formulated within the research. There is no need to spend a long time on carefully crafted research questions before you enter the field. You can start with a "what is going on" type question in relation to your broad interest. This will not make your inquiry aimless. Think of ethnographic research as a funnel (Agar, 1980) and think of "foreshadowed problems" as the starting point.

Foreshadowed Problems

The famous anthropologist, Malinowski, writes that the ethnographer begins his or her inquiry with a "foreshadowed problem" (in Hammersley and Atkinson, 1983: 29). Malinowski distinguishes this problem from rigid pre-conceived ideas that shut the mind to inquiry. In other words, have some sense of direction but be open to shifting this in the light of what you see or hear. In the first phase of fieldwork, the foreshadowed problem, write Hammersley and Atkinson (1983: 39) develops into a "set of questions to which a theoretical answer can be given." This development occurs through a process of focussing as the ethnographer gains confidence in what appears to be noteworthy and important.

To give an example, in his seminal ethnographic study *Learning to Labor*, Willis (1977) started with the foreshadowed problem of what role might working class kids play in their own academic failure. In the course of his exploration in the field, he developed particular lines of inquiry (a set of questions) which were suggested by the boys' apparent macho behavior. These lines of inquiry led him towards a gendered analysis of academic failure and this constituted the theoretical outcome of his research. Willis did not enter the research field with the hypothesis that gendered behavior influences achievement, rather the ethnographic research process brought him to it inductively. Alongside his observations, Willis' sense-making was much supported by his reading—a look at the very full footnotes which accompany his analysis will make that clear.

Reading

Good research is never just about gathering empirical evidence, whatever its form (incidentally, this simple truth gets lost in the empiricist clamour among policy makers for "evidence-based" practice). Delamont (2002) is rightly insistent that ethnographers must read widely so that their explorations and theorizing are always an interplay between what they make of what they see or hear *and* what they read. Similarly, Hammersley and Atkinson (1983) write of the importance of ethnographers taking a wide, comparative view of what reading might be relevant.

Triangulation

Because ethnographers try to attend to different data sources, there is a sense in which they naturally triangulate their evidence insofar as they often bring more than one data set to the table to substantiate their claims. It has become fashionable to reject triangulation on the grounds that each data set might be *erroneously* converging towards the same point, much like different witnesses to an accident might offer partial evidence that results in the wrongful prosecution of an innocent driver. But this reservation should not prevent us from comparing accounts and critically considering the explanatory capacity of different techniques and different kinds of evidence. Hammersley and Atkinson (1983: 199) nicely express the strengths and limits of triangulation:

> Triangulation is not the combination of different kinds of data *per se*, but rather an attempt to relate different sorts of data in such a way as to counteract various possible threats to the validity of our analysis . . . one should not adopt a naively "optimistic" view that the aggregation of data from different sources will unproblematically add up to produce a more complex picture . . . differences between sets or types of data may be just as important and illuminating.

Hammersley and Atkinson (1983: 198–199) also point out that the ethnographer can triangulate his or her theoretical perspectives by comparing them with other researchers' elaborations of a similar issue.

Methods

> There is no codified body of procedures that will tell someone how to produce a perceptive, insightful, or illuminating study of the world.
>
> (Eisner, 1991: 169)

This statement applies particularly to ethnography because it is so much more than a set of procedures or techniques. Indeed, Agar (1980: 114) talks of the "right-hemisphere activity" that is part of the "ethnographic mystique." This does not mean that ethnography is method free—on the contrary—but it does mean that method must be approached with flair and subtlety.

As indicated, reflexivity has to be the constant companion of ethnographic method. The following considerations (applicable to much qualitative research generally) will support the acquisition of a reflexive stance.

1. The Ethical Framework

Your ethical framework and your conduct in the field are inextricably bound. Ethnographers enter settings, producing descriptions and judgments about others. This creates thorny problems of access and consent. Some ethnographers do not secure informed consent purposefully because their research is covert; others secure it at the beginning but it needs to be refreshed during the course of the research. This addresses the fact that once you get to know "informants," they may disclose more than they wanted to. It is of course important to ensure that your interpretations and report do no harm to those affected by the research. To minimize this, do you need to share your provisional analysis with the informants? If not, what defends your decision? If you discover illegal activity, do you report this? Are your informants representative of the group? Did you capture a range of voices and experiences?

2. The Emotional Dimension

This area points up the usefulness of your research diary where you can record how you are feeling in a given situation (uncomfortable, anxious, surprised, afraid, etc.), reflecting on the bearing this might have on data collection and analysis. For instance, if you find a setting gloomy and want to get the observation over with so that you can return home, you might cut short your conversations with informants. Your desire to leave might also give you a sense of how the people who have to work in the setting feel. Your record of this feeling can prompt a line of inquiry as well as an admission that you need to cultivate a higher tolerance of gloom in order to extend your conversations.

3. Producing a Trustworthy Account?

Given that ethnography is about giving an account of what you have found in the field (Agar, 1980), how have you maximized the trustworthiness of your account? What is the quality of your evidence? Have you shared your provisional analyses with other researchers and/or informants? Are you forcing your narratives to support your emerging theory or are you acknowledging the messiness of the field (Law, 2004)?

4. Being Open About Rival Explanations

It is a given in all research designs that we find evidence where we seek it. Can you be sure that you have resisted premature closure of your findings? Have you considered rival explanations to discuss with informants?

5. Researcher Positionality

How informants relate to the ethnographer will have much to do with their perceptions of him or her and the researcher stance they adopt; these perceptions will link to the posture of the ethnographer (e.g. expert, "one of them," undercover member, feigned ignorance, friend, professor-like), his or her physical stature, sex, ethnicity, age, dress, bearing and general impression management. For instance, Paul Willis (1977) is a tall, well built male academic who researched school boys primarily in their schools. Towards the end of his research, the boys confessed to Willis that they started to tire of his presence. Did this influence the answers they gave to him? Their outward behavior? There is likely to be some kind of "ethnographer effect" in the field and this will be shaped by the informants' own impression management as much as the ethnographer's. They might tell you what they think you want to hear. Incidentally, the covert researcher does not overcome this problem because he or she cannot openly discuss formative or final interpretations.

6. Reflexive Distancing

A much debated concern is that of the danger of ethnographers' over-identification with the group they are researching. The risk is that certain behaviors are accommodated rather than critiqued; or the accounts of informants are taken at face value and the ethnographer comes to romanticize and homogenize the community. As Davies (1999: 73) has argued, the quality of research should not rest, as it once did, on how immersed in the field are the researchers because, above all, they need to be:

> sensitive to the nature of, and conditions governing, their own participation as a part of their developing understanding of the people they study.

See Alison Lurie's *Imaginary Friends* (1978) for an amusing spoof of the difficulties of both covert research and of over-identification (sometimes called "going native") with the informants. Reflexive distancing is about guarding against seeing what we want to see.

7. Listening and Learning

Is your posture that of the learner and the listener? This is particularly important at the beginning of fieldwork where the cues about where to focus and what is happening need to be picked up (Agar, 1980).

Mindfulness of the above dimensions will also support your decisions about what kind of posture you want to cultivate in relation to observation.

Observation

The key issues attending observational methods relate to the how and the what.

1. How to Observe

Generally observation methods are presented on a continuum with some taking a position between these:

- *Covert participant observation* in which the ethnographer masquerades as one of the group or *is* one of the group and conceals the fact that he or she is doing an ethnography of it. This posture is regarded as ethically dubious by many researchers.
- *Overt participant observation* in which the ethnographer has negotiated access to the group and participates in its activities as a temporary guest.
- *Overt observer* in which the ethnographer has negotiated access to the group to observe it.
- *Covert observer* where the ethnographer watches a group whose members are unaware of his or her presence. Think of a police stake-out.

Helpfully, Hammersley and Atkinson (1983) argue that these categories all blur into that of participant observation because no ethnographer can achieve complete detachment from the studied group. The ethnographer is always in some sense participating although these categories do allow some thinking about the logistics of fieldwork such as when and how field notes will be written up. Most ethnographers try to write their notes as soon after observation as is practical though the difficulties of accurate recall continues to trouble many as an impossible ideal. In the case of covert observation, particularly, there is no shortage of stories among ethnographers of informants' misinterpretation of the ethnographer's frequent dash to toilets or behind bushes.

Unless the research is completely covert, ethnographic researchers tend to establish a dynamic between observing a group and eliciting accounts from a selection of informants through follow-up conversations and informal interviews. Indeed in Agar's view (1980: 109), where possible, observation should not be treated as the privileged method of ethnography because "observation and interviewing mutually interact with each other, either simultaneously or sequentially in the course of doing ethnography."

2. What to Observe

Broadly the observer needs to pay attention to the routine and ordinary as well as to critical events. Here is an outline of the kind of things to observe and note under the headings of "the environment," "the population" and the "interactive order":

The environment   This might include:

- the architectural layout and state of repair;
- other relevant outdoor features of the setting;
- the smell and feel of the setting;
- the furniture and decor; what is on the walls?;
- graffiti in the toilets or etched into desks; and
- refreshment arrangements.

The population   This is likely to be simple numerical information concerning:

- the number of people present;
- gender, ethnic, age, occupational category and other relevant breakdown of the population.

The interactive order   This is at the heart of your enquiry because it is about how people are behaving in a setting and the meanings underlying their behavior.

Patterned data can support an initial description of the interactive order. You can plot numerically who says what to whom over a given time span (say half an hour of a seminar over a number of weeks). A table setting this out provides a very handy way of discovering at first glance where the interaction appears to take place in a setting. Sociograms which graphically depict the links people have with each other might be useful as well. Generally, the following are some of the things to look for in an interaction order, though of course, the particular field you are researching will determine your own list:

- Arrival and departure of members.
- Who says what to whom, when and how frequently (through both verbal and non-verbal communication)?
- What appears to be amusing? Upsetting? Irritating? Etc.
- How many people are asleep? Eating? Working? Concentrating?
- The distribution of the population (who is where and with whom).
- What roles (formal and informal) do people seem to be assuming?
- What is the jargon of the group (do they have their own terms for people or activities, e.g. swots, partygoers, etc.)?
- Where does the power appear to be? Is there an "in" group and an "out" one? What characterizes them?

While making and recording observations, you will want to formulate questions to ask of particular people or to note where you think you need to focus. Look out for potentially illuminating incidents.

Critical Incidents and Vignettes

Ethnographic reports are much enriched by the provision of vignettes that capture a particular event or activity from which insights can be drawn. For instance, I once observed a flamboyant academic (now retired) strip to his underpants to illustrate a point he was making to his students about the social management of embarrassment. This was a critical incident; such an incident is anything occurring that is out of the ordinary, disturbing or unexpected. It might, for instance, concern the effects of a lawn being mowed outside a classroom; a joke cracked; a group of students complaining and so forth.

Note down both the event you have observed and the reactions of all within its range (including your own). You may well want to follow up your interpretation of the event with questions to test your understanding. In my observation, I wondered whether the students would find being taught by an academic in his underpants quite discomforting but when I probed this with a group of them after their lecture, they all offered assenting nods to the student who replied: "no, no, we love him, it's John, he really cares about us." These incidents often offer promising leads. Indeed, the above example led me to pursue connections between learner achievement and the care relation between academics and students.

Sampling Events

The observer might choose to look at similar events over time to get a handle on how a group behave and the values underpinning the behavior through the lens of such events. Agar (1980: 125) offers the example of weddings in a particular community. In higher education, events for attention might be field trips, graduation ceremonies, induction weeks and so forth. A study of these "high" moments can shed light on the cultural meanings and values of a group.

Videotapes

Where possible, a videotaped record of an event can be an invaluable data source, particularly since it allows the researcher to revisit the site again and again though, of course, it will be a partial view that depends on where the camera pointed—as with the eye in live observation.

Interviewing

Ethnographic interviewing tends to be opportunist and informal in that the ethnographer is often alerted to the need to ask questions as events and behavior occurs. Interviewing is invariably situated in ethnographic research and it resembles more a conversation than a formal event except that the researchers tend to tape record or note down their questions and responses.

Most ethnographers place a great deal of emphasis on rapport building with informants in order to secure high levels of disclosure and reliable accounts (Spradley 1979: 78). Agar (1980) talks of the importance of loosening up in the setting, not diving straight in with questions, but getting a feel for the context and cultivating a good ear and eye as to what is going on first.

Who to Interview

As indicated, ethnographers rely on informants in the field to tell them stories, feedback on interpretations and to answer questions. Agar (1980: 85) writes that most communities will have a "professional stranger handler"; in schools this might be a reliable pupil who is assigned the task of taking parents round the school; similarly, quality assurance auditors in universities might be introduced to particular kinds of staff and students. Professional stranger handlers are very useful; sometimes they are also gatekeepers, determining who else you can speak to. Clearly, research that is solely based on the professional stranger handler is going to be limited.

Ideally, the ethnographer needs to speak to as many informants as possible, particularly if he or she wants to demonstrate the representativeness of views expressed and the reliability of his or her account. Many ethnographers advise the strategy of "theoretical sampling" (Glaser and Strauss, 1967) and here is a good description of it from Agar (1980: 124):

> Theoretical sampling simply means that the ethnographer chooses her next sample to talk to in a self-conscious way to obtain data for comparison with what she already has. Say, you've worked with four men on agriculture. You've talked with them about their interpretation of the flow of events that constitutes agricultural work and you've made several observations working with them in the fields. Now you seek out four more men for shorter interviews/observations who live on the other side of the village. You select them for the purpose of checking similarities in the accounts given by your original sample.

Glaser and Strauss (1967) recommend that the researcher can stop seeking out further informants to talk to once they achieve "saturation" which denotes the point at which they feel they are not learning anything new. Do not seek out saturation for neatness of narrative. However, you might find that there are insufficient "similarities in the accounts" in which case, you can represent different voices in your report (polyvocal accounts).

What to Ask

In most interview methods, the researcher prepares a set of questions before meeting the interviewees. In ethnography, the researcher needs "a repertoire of question asking strategies not a list of questions" (Agar, 1980: 90). These strategies, writes Agar, involve knowing how and when to warm up, nudge, bait, suggest, probe, provoke, declare, encourage, affirm, seek elaboration, request an example, agree, disagree, suggest and so forth.

Below I have paraphrased a selection of strategies from Spradley's 1979: 59, 60) useful list (for a fuller discussion, I strongly recommend Spradley's very accessible book on ethnographic interviewing):

a) The ethnographer offers explanations to the informant as the basis of discussion. In the process, both ethnographer and informant swap teacher–learner roles as they discuss these explanations.

b) The ethnographer makes "I am interested in" statements to keep the discussion centered on the research topic.

c) The ethnographer asks for insider language, asking questions such as "if you were talking to a biologist what would you say?"

d) Structural questions are about the "basic units in an informants' cultural knowledge" (p.60), e.g. "what are the important concepts students have to learn in their first year?"

e) Contrast questions would ask something like "what is the difference between an outstanding student and a mediocre one?"

Another helpful list from Spradley (1979) covers "grand tour," "example," "experience," "insider-language," hypothetical and typical questions:

a) Grand tour—these are "can you take me through what you did in your first month" type questions to enable the ethnographer to get a broad picture (Spradley offers many further subcategories).

b) Example questions ask questions such as "can you give me an example of when that has happened?"

c) Experience questions would ask "do you have any experiences of particularly successful learning in this area?"

d) Insider questions are within the insider's language "so when did you say you felt like you were really at uni?"

e) Hypothetical-interaction questions—"if you got a poor mark from another department, would you still ask about it?"

f) Typical sentence questions ask the informant to say when they use their terms—"so would you say 'uni' to your parents?"

Do not worry about asking leading questions. As you can see, this is not an ethnographer's worry. Indeed Agar (1980) points out that all questions lead somewhere and this undermines the very idea of a non-leading question. Sometimes you will approach people to clear time to ask your questions; other times they will occur to you in the course of conversation. Thus you will not always be able to record interviews and you will need to write up important conversations as soon as you can.

Documents

All documents are cultural artefacts. The ethnographer uses social documents such as institutional records (in universities there are many) not only for the explicit data they may give—which will be of varying reliability—but as "social documents" in their own right. Given that most modern settings are literate, argue Hammersley and Atkinson (1983: 142–143), attention to documents is vital:

> The presence and significance of documentary products provides the ethnographer with a rich vein of analytic topics, as well as a valuable

source of information. Such topics include: how are documents written? How are they read? Who writes them? Who reads them? For what purposes? On what occasions? With what outcomes? What is recorded? What is omitted? What is taken for granted? What does the writer seem to take for granted about the reader(s)? What do readers need to know in order to make sense of them?

Other topics might concern the identification of attributes of, say, "good teaching" in quality assurance reports or of the "good" exam script. Others might concern an exploration of confessional narratives from teacher development portfolios or the tone and language used for guidance documents to students on plagiarism.

Diaries from Informants

Some ethnographic data can be generated by informants themselves if they agree to keep a diary, either written or videoed. They will need some guidance as to what you are looking for and how much feedback and reflection to provide.

Field Notes: Data Collection and Analysis

As indicated, data collection, data analysis and the writing of the research report are not neat sequential activities but dynamically linked in ethnography. Field notes are the expression of this linkage. They capture observational and interview data, interpretations, ideas, theoretical elaboration, possible typologies, thematic links and cross references with the research diary. They can include photos, sociograms, drawings and discussions of informant diaries. They are, writes Delamont (2000: 64), liminal texts because:

> They are on the borderline between "the field" and "home," between "data" and "results," between "private" and "public" records. Field notes are not a closed, completed, final text, rather, they are indeterminate, subject to reading, rereading, coding, recording, interpreting, reinterpreting.

And Sandra Kouritzin (2002: 133) refers to them as half-baked data because:

Every time researchers choose a word, or a sentence structure, every time they use active or passive voice, or direct or indirect reporting of speech, every time they use a known narrative structure, researchers create the very evidence they will later use as "proof" of their interpretations. The concept of "raw data" is therefore "half-baked."

Finally, Agar (1980) refers to them as "working notes." He cautions against writing so many that they form a daunting, unstructured pile on the researcher's desk or in the computer. Hammersley and Atkinson (1983: 165) advise that the best way of avoiding data overload is the frequent review of field notes for the writing of analytic memos:

> Regularly review the development of analytic ideas in the form of analytic memos. These are not fully developed working papers, but periodic written notes whereby progress is assessed, emergent ideas are identified, research strategy is sketched out, and so on.

When you write field notes or transcribe interviews, leave lots of space on the right hand side to support coding and thinking. Short analytic memos can be written in different pen colors or at the margins of the data; longer memos will have to be clearly tagged to relevant data. I strongly recommend that you store the field notes and transcripts in software (e.g. Atlasti or NVivo) if you can because this allows for both effective data management, analysis and memo-ing.

Unless you particularly love computers, do not sign up to training that takes you through every bell and whistle enabled by the software; it will overwhelm you and you are likely to forget most of what you learn. Just go for mastery of the essentials and then practice. Within a week, I learnt enough to get going, namely to prepare data for the software, to codify the data and to write memos. If, like myself, you like to work with paper, you simply print out various data sets and memos to explore them afresh.

It is helpful to recall the funnel metaphor again because field notes support the process of progressive focussing. The researcher, writes Delamont (2002: 170) should return to her field notes, to:

> read/re-read field notes/diaries—draw out both recurrent patterns and instances that run contrary to those patterns. Themes and categorisations are extracted during these recurrent readings.

For those who are unsure about how to write up field notes—or indeed what to look for in the early days, Agar (1980: 92) suggests following the journalist rule in which an account incorporates the 5 "W's,"namely:

- *Who* was there.
- *What* happened and what was said.
- *When* did it happen.
- *Where* did it happen.
- *Why* did it happen (this might involve various explanations from informants).
- *How* did it happen (again this might involve various explanations from informants).

Some of the field notes you write will express "experience-near" concepts, that is those generated by informants about the world in which they move; others will be experience-distant and have more to do with the researchers' interpretations and their world (Geertz 1973; Agar 1980). These are important distinctions for ethnographers who are on the look out for "folk taxonomies" from experience-near data. For instance, students might be heard referring to "slackers," "swots," "partygoers" and "last-minuters" to describe different kinds of students. Experience distant taxonomies would be generated by the researcher such as those of "strategic," "deep" and "surface" orientations to learning.

There is a constant movement in ethnographic research between gathering the data and making sense of it. See Chapter 3 for further guidance on how you might approach continual analysis of the data. See also Grbich's (2007) lucid account of a number of analytical moves that can be made from classical and critical ethnography.

Writing Up

Much ethnographic writing aims for the provision of a thick description. The "description" part of this term can be misleading because it has to be thick with the analytical as much as the descriptive. The ethnographer, explains Delamont (2002: 170), aims to produce:

> An account of the culture or institution being studied which would enable the reader to live in it without violating its rules.

Some of the rules such as "do not walk on the grass" are easily described and comprehended but more complex ones such as those governing promotion to professorial status require a deeper explanation of what that involves, to embrace, for instance, formal and informal rites of passage, cultural signs of membership, exclusive practices and so forth. It is worth going back to Geertz's (1973) discussion of this topic in his very interesting essay *The interpretation of culture*.

Clifford Geertz (1973) offers a playful discussion of the proposal from the philosopher Gilbert Ryle that the mere observation of a wink cannot tell you whether it is a physical reflex or a form of communication; and if a form of communication, continued Ryle, its symbolic meaning given the cultural context in which it occurs is not self-evident and even if it were, how can an observer know that the wink is not a satirical subversion of its usual cultural meaning? In the face of this kind of interpretive predicament, which involves layers of possible meaning, argued Geertz, we must avoid the "thin description" of behaviorism which would simply record that a wink had taken place. Instead we must aim for a "thick" description which will strive to grapple with complex, multi-layered meanings, so that winks can be distinguished from "twitches and real winks from mimicked ones" (Geertz, 1973: 5).

Thus a thick description which fulfils this aim would have to do more than offer a description of people winking at each other. The account would have to say why and when they appear to be winking. If the analysis of the data has led the ethnographer to the view that the winking is part of a courtship ritual, then an illustrative and analytical vignette of a winking encounter would bring this home to the readers, both allowing them "to be there" and to know when and with whom it is appropriate to wink in the group studied.

Your writing, then, must be analytical. Thick description, as Shank (2002: 75) reminds us is not "voluminous description." Firstly, points out Shank, the detail it includes has to be meaningful for the theorizing in hand; if the fact that two students wore blue fleeces is irrelevant, do not mention it. On the other hand, if the fleeces denote membership of a football club and this is important for the analysis, it needs to be included. Secondly, the thick description cannot be "idiosyncratic" (Shank, 2002: 76) since it has to have some linkage with a wider frame of reference. For instance, my thick description of the disrobing academic discussed above needs to address an awareness that this is quite transgressive behavior even in a culture (British academe) where eccentricity is tolerated. It needs also to address the unexpected finding that the students did not find this so because of their overriding trust and respect for this teacher. In sum, in Shank's words (2002: 77) "the task of thick description is to make meaning clear."

Conclusion

Higher education researchers are unlikely to have the conditions for a full blown ethnographic research project. The aim of this chapter has been to introduce this important tradition from naturalistic inquiry to enable the reader to use some of its insights, ideas and methods.

Further Reading

Agar, M. (1980) *The professional stranger*. New York: Academic Press.

Davies, A.C. (1999) *Reflexive ethnography: A guide to researching selves and others*. London: Routledge.

Delamont, S. (2002) *Fieldwork in educational settings: Methods, pitfalls and perspectives*. London: Routledge.

Grbich, C. (2007) *Qualitative data analysis: An introduction* (Chapter 2). Thousand Oaks, CA: Sage.

Hammersley, M. and Atkinson, P. (1983) *Ethnography principles in practice*. London: Routledge.

Shank, G.D. (2002) *Qualitative research: A personal skills approach*. Upper Saddle River, NJ: Merrill Prentice Hall.

Spradley, J.P. (1979) *The ethnographic interview*. Fort Worth, TX, and London: Harcourt Brace Jovanovich.

8

CASE STUDY RESEARCH

The Appeal

Case study research is well suited to inquiries into "processes and relationships" (Denscombe, 2007: 38) and to broad research questions. Case study researchers "recognize the complexity and 'embedded-ness' of social truths" (Adelman et al., 1980: 59) and the difficulty of capturing these through controlled experiments or statistical analysis. This research approach offers the opportunity to investigate issues where they occur (naturalistic settings) and to produce descriptive and analytical accounts that invite reader judgment about their plausibility.

Purpose

Case Studies and Case Study Research

The generic term "case study" sometimes clutters the meaning of case study research. Instances of "good" or "best practice" gathered for teaching or development purposes should not be confused with case study research. Accounts of the former tend to offer a victory narrative through which the issue or problem is defined and the triumphant solution described. In contrast, case study research systematically explores a setting in order to generate understandings about it. In higher education the case could be anything from a particular initiative to the whole institution.

Naturalistic Inquiry

As a form of naturalistic inquiry, case study research provides a holistic approach to the exploration of real life situations. It involves the gathering of data from a variety of sources and methods (e.g. observation and interviews) within a specific setting. As elaborated below, case study research may compare a range of cases or it may focus on one single case for local and/or general understandings. Either way, the research always takes place where the case naturally arises; it is research of an "instance in action." The people in the case study are best described as "actors" (rather than subjects or informants) to capture this.

The Case as the Focus

Case study research starts with a research curiosity about a particular case, asking what is going on with this person, institution, program, etc. Case study research is well suited to evaluation research, and for this purpose it will combine curiosity about the case with an eye on the declared aims of the activities being evaluated.

Theoretical Concerns

Researcher Position

Case study researchers range from adherents to logical positivism (Yin, 1993, 2002) through to various post-positivist frameworks (Simons, 1980; Stake, 1995; Bassey, 1999), This chapter draws heavily on the work of Robert Stake for three reasons: firstly, Stake provides a very accessible navigation through the various positions of positivists and post positivists to a landing place that acknowledges: a) the influence of an external reality on our meaning making; b) an acceptance of the mediated character of our meaning making; and c) the possibility for research to produce "integrated interpretations" (Stake, 1995: 100) from a) and b). In other words, the world may not be entirely knowable but we can know something of it. While Stake accepts that our interpretations will always be provisional, he does not see them as hopelessly arbitrary; rather they are produced from evidence-informed "good thinking" within scholarly communities.

This position—to which we can add a post-modern concern for language (see Chapter 1 for a discussion of this)—provides a basis for an accessible, contemporary, interpretivist framework well suited to case study research.

Secondly, Stake's approach suggests a respect for techniques of data gathering and analysis that does not tip into technicism. Essentially, advises Stake, you must "place your best intellect into the thick of what is going on" (Stake, 2000: 445). Rigor from this viewpoint is achieved through thoughtful, scholarly engagement with empirical data. Thirdly, Stake's discussion is rooted in the research and evaluation of education settings and thus of central relevance to this book.

Types of Cases

Stake (1995) offers the following three categories of case study research:

1. *Intrinsic case study*   This is where the researcher's interest is in understanding the particularities of the case in hand. Intrinsic case study primarily aims to generalize *within* the case.
2. *Instrumental case study*   This is where the researcher explores a case as an instance (e.g. a specific geography field trip) of a class (e.g. geography field trips) in order to shed light on an issue concerning the class: e.g. what is happening on this geography field trip that can tell us something about geography field trips in general? The next category, collective case study extends this kind of generalization.
3. *Collective case study*   This is where researchers select more than one case of the class from which to generalize: e.g. what is happening on these field trips that can tell us something about field trips in general?

What type of case study you choose will partly depend on logistics of access and resources available and partly on your aims. Collective case study research is often conducted by a team of researchers who agree a framework for data collection and analysis across similar settings. If this is to work well, the team needs to meet regularly to compare and explore findings and to agree ongoing foci. Intrinsic case study research is well suited to evaluation research.

Petite and Grand Generalizations

Case study researchers should not get hung up on making generalizations from their research. Simons (1996), for instance, argues that a drift towards multi-site case studies risks neglecting this research approach as essentially a "science of the singular" which aims to grapple with the complexity, depth and uniqueness of a case and to self-consciously dance with the tension between the particular and the general.

Stake is in broad agreement with Simons' point but he writes that forms of "petite generalization" can be an important element of single case study research. For instance, it might be discovered that when students in an intrinsic case are asked to form discussion groups, they repeatedly do so in particular ways; a generalization can be made about this repetition. Alternatively, a case study might confirm or contradict a pre-existing "grand" generalization. If, for instance, the literature says that students always group together according to gender and a case study finds that this is not so, then it potentially modifies this generalization. Incidentally, Stake prefers the term "assertion" to that of generalization. He also counsels the researcher to adopt an "ethic of caution" (1995: 12) when making assertions.

Fuzzy Generalizations

Bassey (1998, 1999) comes at the issue of generalization from a complementary place to that of Stake's. In Bassey's (1998) view, the most appropriate aim for case study research is to make "fuzzy generalizations." At a conference address, he engagingly explains how he came to this position:

> I was struggling to find a way of expressing succinctly the idea of a qualified generalization when I came across a paper by C. Fourali called "Using fuzzy logic in educational measurement." (Fourali 1997) This resolved for me a problem with which I have often grappled as an examiner of student papers. Instead of trying to give an exact mark— like 57 out of 100 for an essay, Fourali advocated giving a fuzzy mark, like 50–60 out of 100. If another examiner gave a fuzzy mark of 55–70, then it might be appropriate to combine the two and give a narrower

range of 55–60 as the moderated mark. Then it dawned on me that this was what I was looking for: my "qualified" generalization could be described as a "fuzzy" generalization.

Translated into research practice, Bassey suggests that we should tone down any talk of probability by using "may" rather than "will": e.g. our case studies show that if students do not undertake fieldwork, they *may* have difficulties in learning to think like geographers.

Naturalistic Generalizations

Finally, returning to Stake, there is a twist to his argument. He writes that "the real business of case study is particularization, not generalization," adding:

> We take a particular case and come to know it well, not primarily as to how it is different from others but what it is, what it does. There is emphasis on uniqueness and that implies knowledge of others that the case is different from, but the first emphasis is on understanding the case itself.
>
> (Stake, 1995: 8)

And yet, it is this very attention to the depiction and analysis of the uniqueness of a case that allows for a form of generalization to be made, not by the researcher but by his readers:

> People can learn much that is general from single cases. They do that partly because they are familiar with other cases and they add this one in, thus making a slightly new group from which to generalize, a new opportunity to modify old generalizations.
>
> (Stake, 1995: 85)

A key aim of case study research be it intrinsic, instrumental or collective is to offer a wealth of readable detail and analysis, such that the reader can make a judgment about the case. This judgment is a "naturalistic generalization," involving:

> conclusions arrived at through personal engagement in life's affairs or by vicarious experience so well constructed that the person feels as if it happened to themselves.
>
> (1995: 85)

Enabling such naturalistic generalizations through skilful, thick descriptive, write-up is a key to successful case study research, of which more below.

Triangulation

A number of case study researchers, including Stake (1995), invoke Denzin's (1984) four types of triangulation as a way of ensuring that the case study is reliable—these are:

Data source triangulation   This involves exploring the effects of context; it is "an effort to see if what we are observing and reporting carries the same meaning when found under different circumstances" (Stake, 1995: 113). For instance, if students in a particular lecture series are observed to be inattentive, then watching their behavior in a different series with a different teacher, in a seminar and in peer learning groups will allow the researcher to see if the inattentiveness is a constant across these settings.

Investigator triangulation   This is where other researchers are asked to make similar observations or look at the same data to "support or undercut the original interpretation" (Stake, 1995: 113).

Theory triangulation   This is where researchers from alternative theoretical viewpoints are asked to look at the same data to offer rival explanations.

Methodological triangulation   This is what most people take to mean triangulation because it involves bringing different methods and data sources to bear on a provisional finding. Many see this aspect of triangulation as involving a mix of quantitative and qualitative data (often called "mixed method") though the mix can be of any approach.

Since Denzin set out his protocols for triangulation, there has been growing acknowledgment that triangulation affords no way of knowing whether:

agreement between the results of two methods "proves" the validity of the second method as well as the first (the principle of mutual confirmation, also known as "arguing in a circle").

(Massey, 1999: 183)

This kind of scepticism about aspects of triangulation need not lead to its entire abandonment. In particular, the rejection of triangulation should not become a licence to be less attentive to the quality of data and argument. Broadly, whether you call this triangulation or simply good research, it is best to get others to look at your interpretations of the data, to watch what is happening across different contexts, to identify any tensions in diverse data sources and to offer any additional interpretations that arise. You can add to this process a degree of reflexivity about rival explanations to your analysis.

Method

Messy Terrain

It is hard to be prescriptive about the design, implementation and analysis of case study research because it is a messy business, involving a degree of connoisseurship (Eisner, 1991), that is, a nose for an emerging focus, a supportive theoretical literature, exemplary stories and vignettes, appropriate methods to use, analytical moves to make, data to shed or to keep and write-up flair. There is some overlap between the framework for case study research presented here and that of ethnographic field work. In many ways, this chapter and the one on ethnography are best read together.

Reconnaissance

As with any research method, case study research requires an engagement with a promising literature and available documents from the case to support the researcher's bearings. This is not about undertaking a literature review before entering the setting, rather it concerns stimulating the formulation of research questions for the beginning of the study; it also concerns securing a continual engagement with theory throughout the empirical research process.

Formulating Research Questions

Given the goal of case study research to explore a setting for what it throws up, hypothesis led research is not appropriate to this method. Yin (2002) suggests that case study research largely require "why and how" questions because they invite an investigation into meanings and explanations. How far these can be planned ahead of entering the case is partly a matter of choice.

Preparing "Issue Questions"

Stake (1995: 25) suggests that the researcher plans a number of "issue" questions that prompt "good thinking"; these might have a cause and effect nature, such as:

> Do students who undertake field work have an attainment edge on students who do not do field work?

Or they might center on a problem or a puzzle (perhaps inspired by a reading of the literature):

> Are the international students getting less out of the fieldwork experience than the home students?

Further questions will "call for information needed for a description of the case" (Stake, 1995: 25):

> What preparation for the field trip do the students get?

Some questions lend themselves to the gathering of coded, numerical data because they concern variation, association and frequency:

> How many times do teachers and students respectively talk?

Some might concern espoused theory and its practical translation:

> How are the declared aims of the programme expressed in the documents? Are there local interpretations of this?

Most of the questions are likely to concern the setting itself but a few will relate to the researcher positioning (e.g. am I seeing things through the lens of my own experience?) and outside influences (e.g.

are the aims of the programme primarily written for a quality assurance audience?).

Stake suggests that you are likely to have around ten to twenty questions at a first stab and that these will be reduced to three or four as you refine your focus. In doing so, bear in mind that you need to be confident that the questions can be answered in the research setting (Gillham, 2000: 16). Keep a note of the questions you formulate as you go through the research process.

Research Diary

Because of the complexity of case study research settings, it is advisable to start a research diary for record keeping, contact details, appointments, tasks to undertake and personal reflections. Stake's (1995: 51) suggestion is that you draw up a data gathering plan with the following elements as an early entry:

> Definition of case, list of research questions, identification of helpers, data sources, allocation of time, expenses, intended reporting.

Your choice of the methods of data collection (discussed below) is clearly going to be linked to what seems appropriate and feasible in the setting and, what might respond to your questions. Be prepared to modify your plans about what and how to collect as you enter the setting and discover both unexpected opportunities and blocks.

Selecting the Case

Your selection of a case or of cases will of course depend upon what you want to explore. Denscombe (2007: 40) suggests four different kinds of cases, namely:

> *Typical instance*: a case that seems typical of others in other settings.
> *Extreme instance*: to explore how this might contrast with the typical.
> *Test-site for theory*: the case would be explored to see if it bears out an existing theory.
> *Least-likely instance*: this tests a theory by exploring whether it holds good in an unlikely setting.

One could add to this list Yin's *exemplary case* which can "reflect strong, positive examples of the phenomenon of interest" (Yin, 1993: 12).

In making your selection, access issues will be paramount. You may want to research an "extreme instance" but only have good access to a "typical one."

Ethical Access

Ethical clearance will be important of course. Everyone within range of your study will need to know that you are exploring the case and a written summary of the research aims needs to be made available to them. The best way of securing consent is to involve subjects in your reflections and lines of inquiry throughout the research cycle. Demonstrably take up the posture of a learner and be mindful of the rights to privacy of the actors.

Sample

If you are going for cross-site comparisons, it is important to note that although you may strive for representative research sites, case study research is not concerned with a strict process of "sampling" for a representative set of cases. Case study research aims for depth and if a site cannot deliver this due to limited access to actors, events and settings, it will not yield the data needed for the study to be successful. "If the sampling logic is important to an inquiry" writes Yin (1993: 34), "the survey or experimental methods are more likely to satisfy an investigation's needs."

Defining the Case

The case is your unit of analysis so you need a clear definition: in medicine, the case is often a patient; in business studies, a workplace; and in education research it can be a programme, a single institution (school, university), a course, an individual student, a group of students and so forth.

Defining the Boundaries of the Case

An important early task is for the researcher to define the boundaries of the case by making decisions about the following dimensions:

- *Physical borders* Are you researching the biology student experience within the confines of teaching and learning activities in the department? Will you include a field trip or a social activity within the case?
- *Population* Who are of concern to your case: anybody passing through the research site? Students? Academics? Cleaners?
- *Range of activities* Is your case limited to the students' teaching and learning activities in seminars and lectures? Or will it extend to their "down time"?
- *Time span* Will you explore the students' experiences over one semester? Will it involve specific slices of their day over this period?

As you research your case, you might well redraw the boundaries in the light of emerging evidence but an initial working definition needs to go into the research diary.

Data Collection and Analysis

Case study research is rarely a single method approach and usually involves any combination of the following ways of gathering and analysing data (note again the similarities with ethnographic research); as indicated, some of these methods you will have planned in advance, others become available or important during the course of the research.

For broad issues of qualitative data analysis see Chapter 3. For the purposes of this chapter, I have collapsed collection and analysis together because they are deeply intertwined in case study research; "analysis" writes Stake, "is a matter of giving meaning to first impressions as well as to final compilations" (1995: 71). The continual meaning making thrust of case study research also allows *in situ* judgments to be made about what kinds of data needs to be gathered.

Documentary Analysis

This can be carried out during the reconnaissance phase though it will be ongoing involving any documents (e.g. minutes of meetings, programme documentation) for which you have consent to read. This can also include visual documents—photos, pictures, etc.

Interviews (Group or Individual)

These might initially be information seeking, though you will also want to discuss your emerging ideas and elicit perceptions, perspectives and experiences. Where the interviews occur in a moment of happenstance, they are more likely to resemble conversations in which you take a gentle steer. For this reason—and those of economy—you are more likely to take notes than to tape record the event.

Critical Incidents

Critical incidents might be observed and recorded by yourself or you might ask subjects to provide them. In a classroom observation, you might note, for instance, how the students reacted when the fire alarm was unexpectedly sounded; how many students returned to the room? How did the teacher settle them again? You might then compare your own impressions with those of the teacher and students by talking to them afterwards. These incidents may provide you with illuminating vignettes about student commitment to their learning. Key questions to ask for gathering critical incidents are:

1. Can you describe an episode that stood out in some way?
2. What happened? Where? When? Who was involved?
3. What part did you play in it? (active, bystander, etc.)
4. What did you make of the episode?
5. Might there be an alternative interpretation?

Stories and Life Histories

Critical incident narrations often have a story structure (a past event which has a beginning, middle and an end) but if your focus is explicitly on gathering stories, you want the narrative structure to come from the teller rather than from structuring questions from yourself. It is partly through our stories that we construct a sense of who we are. This is why the researcher is interested in how a told story offers clues about the teller. Apart from exploring the content, the researcher examines the story for questions such as: has she provided a defensive narrative? Has he provided a self-mocking construction of masculinity? What does his story suggest about the actor's positioning in the team? See Chapter 6 on narrative inquiry.

Life Histories

You might want to collect "life history" data, particularly if the case is an individual or a small group. This will involve decisions about which slice of their life is important: it could be a full length biography, a retrospective account of a specific period or an account of the present. You can provide thematic headings (e.g. your undergraduate years) and ask for written accounts perhaps over a period of time; or you can record life histories within an interview (notes will not do for this kind of research). Ensure that you do not invite one dimensional accounts that support a particular line of inquiry. See Chapter 6 on narrative inquiry.

Numerical and Patterned Data

Be aware of what is countable (e.g. repetitions, frequencies, absences, possible associations) and devise ways of recording this systematically. Stake (1995: 76) stresses that both the unique and the singular as well as numerical and patterned evidence is of value, as he elegantly states:

> the quantitative side of me looked for the emergence of meaning from the repetition of phenomena, the qualitative side of me looked for the emergence of meaning in the single instance.

For Stake (1995: 45), the identification of patterns in qualitative research is the equivalent of "correlation or covariation in quantitative study."

A common strategy for case study researchers is to use a simple grid to codify variable behavior. For instance, teacher talk and student talk can be counted to address questions about classroom interactivity or degrees of teacher-center-ness. The talk can be subdivided into, say:

- positive feedback talk;
- student contributing talk;
- information seeking talk;
- information giving talk; and
- instructional talk from the teacher.

This can be monitored over a specified period of time (e.g. a middle chunk of fifteen minutes of a particular lesson over a period of a week). The researcher simply ticks an appropriate category on the grid every time he or she hears an example of it, cross referencing this with some indication of who provided the example. He or she will end up with a patterned representation of the talk, showing, for instance, who talks the most about what. There are likely to be a range of opportunities to use this technique in any one setting though be careful you do not drift into a simplistic interpretation of what you count or a sense that you are being more "scientific" by using these methods.

Another possible source of patterns might be found in your record of conversations: are there thematic recurrences? Frequent references to time poverty? Research deadlines? Can you give meaning to these frequencies? For instance, can they be associated with a lower priority given to teaching? (This kind of inference can be further explored with those who were party to the conversations.)

Be on the lookout for possible associations. Are students more alert at afternoon lectures than morning lectures? You might design a grid with indicators of degrees of alertness such as "punctuality," "note-taking," "attentive posture," "asleep," "texting," "chatting," "sitting at front" quantitatively assessing these by observation of the first and last ten minutes of the lecture; if there is an unmanageable number of students, you will have to select, say, two middle rows of the lecture room.

Sociograms

If you are exploring a team or group interactivity among students, you might want to consider sociograms; these are graphic mappings of the social links people have. Wikipedia has a useful entry on sociograms with examples and links.

Questionnaire

If your case is a course or program, you might want to distribute a questionnaire to students to support your judgments about how they are experiencing their learning and the teaching. Edinburgh University has a very useful webpage that references and details instruments that can be used for this purpose www.tla.ed.ac.uk/etl/publications.html#measurement.

Observation

There is a full section on observation in the chapter on ethnography and I will make some brief comments to supplement this. Observation is undertaken either as a participant or detached observer. From an ethical viewpoint the observation is usually overt so that actors are aware of your research activities as far as is possible.

Observation is a rich way of exploring human relations and processes (the interactive order) and for depicting the environment. You will have your questions to guide you but it might be helpful to think of observation methods as being close to good detection work. You know what you are looking for but you do not know what you are looking for in relation to what you are looking for! Consider the ways in which a detective meticulously builds up a picture of his case, using whatever observational evidence is to hand, giving weight to what might seem trivial or irrelevant as well as to what might seem of obvious relevance.

You might want to note down jokes told and responses to them; you might want to look out for what seems to prompt the anxieties or the delight of actors, how they work as a team, what kind of roles individuals appear to be taking up and so forth. A readable book to support observation of the interactive order is Goffman's classic (1959) *Presentation of Self in Everyday Life.*

You will need to describe the environment: how is the room arranged? What is the age of the furniture? What have the students etched onto the desks? What notices are on the wall? Observations are recorded in field notes and your reflections and theorizing on them in a research diary. In describing the settings and actors' relationships in your field notes, do not worry if you feel that you are "straying" into a literary genre because some case study research writing self-consciously blurs the boundaries between social science and literature (Geertz, 1983), as I discuss below.

Visual Data

Photos, videos and drawings can all enrich both the analysis and the depiction of the case study. These can be used as a means of explication in their own right and/or as a way of prompting discussion and the development of interpretations.

See Chapter 11 on visual research methods.

Managing the Data and the Analysis

Throughout this book, I recommend using qualitative data analysis software; this recommendation is particularly strong for case study research because it facilitates both data management and analysis. Case study research involves a complexity of sources and data sets and the software will enable both ease of storage and of cross referencing. This does mean that interview notes, field notes and research diary entries will need to be in typed form but the process of typing is not just another chore since it increases your familiarity with the data. You can, of course, just type a shortened version of your notes.

Stake (1995: 85) cautions against accumulating a daunting data mountain, preferring to analyze and shed as he proceeds. While a constant screening process to ensure manageability of data is important, there is a case for swapping the "mountain" imagery of data overload with that of a rhizome. Delueze and Guattari (1987) argue that we are attuned to think through tree-like hierarchies because they express cultural norms about power and importance. An alternative, they argue, would be to structure our thinking through the image of a rhizome. This image facilitates thinking about "additions" to our interpretations as well as about converging or commensurate data. A rhizome is a root (like ginger) which has lots of shoots running off it horizontally. With this image we can also think of the data as spread out rather than piled up; this might better suit the variety of case study data sources, the identification of alternative interpretations, anomalies and conflicting voices.

Here are some questions to support your analysis of the data:

1. Are there some clear themes emerging?
2. Have you got something to claim in the way of fuzzy, petite or grand generalizations?
3. Can you offer good chunks of data to underpin your interpretations?
4. Are there evocative vignettes to write from observations?
5. Is there some interesting patterned data? What data does not fit the pattern? What might it be indicating?
6. How does this/these case studies relate to others? And to the theoretical field?
7. What gaps and puzzles remain?

Writing the analysis

As indicated, the write-up starts from the beginning of the research process. In the initial stages you have formulated a defence for your case selection, the issues to be addressed, the boundaries and the supportive literature. About two-thirds of the writing will have been interwoven into the processes of data collection, data analysis and diary noting.

There is no right way of writing up case study research but it needs to be interesting, compelling and convincing. In particular, it needs to draw in the reader, enabling him or her to vicariously experience the setting and thus to make "naturalistic generalizations." Like ethnographers, case study researchers often rely on the notion of thick description (Geertz, 1973). Such description attempts to present layers of meaning (see the chapter on ethnography for a further discussion). The narration of vignettes in particular can be written within the spirit of the literary turn that characterizes some contemporary social science. See Shank (2002) and Seale (1999) for further discussion of qualitative research writing genres from realist to postmodern styles. You will need to make a decision about whether you present the data first and then discuss your interpretation, or whether you want to integrate the presentation of data into the account. Whichever style or structure you develop, keep in mind that the reader needs to feel that they could "be there" so that they can share in the interpretation with you. I will leave the final word on writing to Stake:

> Where thoughts come from, whence meaning, remains a mystery. The page does not write itself, but by finding, for analysis the right ambience, the right moment, by reading and re-reading the accounts, by deep thinking, then understanding creeps forward and your page is printed.
>
> (Stake, 1995: 73)

Conclusion

Case study research has the potential to generate rich understandings be they of a single case or a set of similar cases; it offers flexible and creative ways of researching live settings; and it licenses evocative write-ups that aim to describe, interpret and persuade the reader.

Further Reading

Adelman, C., Jenkins, D., and Kemmis, S. (1980) Rethinking case study: Notes from the second Cambridge conference. In H. Simons (Ed.), *Towards a science of the singular*. Norwich, UK: CARE, University of East Anglia Occasional Publications.

Bassey, M. (1999) *Case study research in educational settings*. Milton Keynes, UK: Open University Press.

Simons, H. (1980) *Towards a science of the singular*. Norwich, UK: CARE, University of East Anglia Occasional Publications.

Stake, R.E .(1995) *The art of case study research*. Thousand Oaks, CA: Sage.

Yin, R.K. (2002) *Case study research, design and methods*, 3rd ed. Newbury Park, CA: Sage.

9

ACTION
RESEARCH

Appeal

Action research offers a means by which research and development
(be it institution wide or at the level of local practice) can be combined
within a framework of public, reflective inquiry. It offers profitable
ways in which groups of academics (or academics and students) can
investigate together issues that might be puzzling, troubling and/or
sensitive to them through a solution-oriented approach. It can also
enable individuals to research their own practice.

Purpose

Scale

There is a common misconception that action research is only suitable
for small scale, practice-based research. This is not the case since
action research can be used to explore anything from assessment
practices on a single course through to a change management project
in a whole institution or across a cluster of institutions.

Different Models

There are a number of models of action research. Given that action
research is used across many subject areas, particularly those of business
studies, community development, education and health, this is hardly
surprising. Roughly speaking, at one end, action research, particularly

in the UK field of schooling research, is associated with the "teacher-as-researcher" movement (Stenhouse, 1975; Elliot, 1991; McNiff, 1994). This orientation can be understood as extending the principles of reflective practice into systematic, public inquiry. In the middle there are versions which use action research to discover "what works" (Zuber-Skerrit (1996) calls this technical action research) and, at the other end, there are models of emancipatory action research (Carr and Kemmis, 1983; Reason and Bradbury, 2001) which aspire to change the world for a more equal and inclusive one. This latter model is often termed PAR, (participative action research) or radical action research. Janet Masters (1995) provides a good history of action research and its various traditions.

Whatever the different models, they all have the purpose of researching a proposed change within everyday, natural contexts rather than within controlled settings.

Theoretical Concerns

Many action research perspectives sit within the interpretivist tradition though they can equally lean towards positivist or radical post-modern frameworks. Whatever the leaning, they all value an iterative, reflective and cyclic process of exploration for the generation of practical outcomes. For some, action research is a handy method for particular problems and for others it is integral to a holistic, ethical way of being in the world.

Under Method I will present a cross-breed based on key principles and procedures from prevalent models of action research from interpretivist to emancipatory traditions. These models are largely about team-based action research though they can be adapted to individual, practitioner-based levels (see McNiff, 1994, for a particularly good guide to practitioner-based action research).

Researching On

Consider the following statement from a sociologist writing in the 1970s:

Different classes in our society not only live differently quantitatively, they live in different styles qualitatively. A sociologist worth his (*sic*) salt . . . can make intelligent guesses about the type of pictures on the wall and books or magazines likely to be found on the shelves of the living room . . . predict the number of children sired by his subject and also whether the latter has sexual relations with his wife with the lights on or off.

(Berger, 1963 in Reid, 1977: 18)

The striking feature of this passage is its confidence in classification and prediction. There is no nuanced, complex, empathetic, differentiated, gendered understanding of working class people here; they are placed in boxes, even with respect to the most intimate aspects of their lives. This is a result of observing subjects at a distance. It is precisely this kind of observing and of the categorizing and "otherizing" that flows from it that action research seeks to overcome. Otherizing involves treating people as fundamentally different to oneself. The presumption that a sociologist "worth his salt" can generalize confidently about large slices of the human population is called into question by action researchers.

Action research won credibility in the aftermath of World War II and in the light of failures of other forms of research to get to the bottom of critical social problems such as that of racism (Lewin, 1946). It was felt by the proponents of action research that bringing social scientists closer to the experiences of the people who had to endure them would yield better knowledge and better solutions. Broad surveys had their place but they were felt to conceal as much as they might reveal if left to fend for themselves as a sole method of discovering key issues associated with particular populations.

Researching With

The important idea for action researchers is to research *with* subjects, not *on* them. This idea reverses the conventional scientific understanding of objectivity. Action researchers argue that distancing the researcher from the researched in order to achieve objectivity merely risks generating knowledge restricted to the researcher's world and prejudices, much like Berger has offered above. We see what we look for and what we look for is often hooked into our own perspective and values, particularly regarding those who are not like "us." Thus what is frequently paraded as "objective" knowledge acquired from a neutral distance may well be objectifying of the human subjects under scrutiny.

From the action researcher viewpoint, then, non-dialogic, non-participative forms of research are seen to run the risk of otherizing research subjects, often reducing them to stereotypical versions of themselves. In contrast, research *with* subjects is held to create a climate of inquiry that is generative of more disclosed, informed, subtle, appreciative, negotiated and intelligent understandings.

While the critique of "researching on" opens up a promising path for collaborative research, it is important to bear in mind that "researching-with" brings its own difficulties. For instance, some action researchers often make very strong claims for the emancipatory function of action research, as in the statement below from Zuber-Skerritt (1996: 12.13):

> [Action research] is emancipatory. The approach is not hierarchical: rather all people concerned are equal "participants" contributing to the enquiry.

The view that there is no principle research team, hierarchy or "outside expert" has to be tempered with realism. There are likely to be substantive inequalities within most collaborative research projects whatever their formal democratic structures. Academics, for instance, have power over students. The best remedy for a sound action research model is to own up to any inequalities in the research group and to address how best to diminish their impact.

Another source of inequality will lie in differentiated investment and yield from the action research. Those initiating (and often paid to do so) the research and perhaps aiming for publication will often put more into the research activity than those less centrally involved (such as students). Thus a good action research model will both value and problematize expert contributions and leadership in the light of diverse levels of preparation and experience among the researcher group (Cousin, 2000; Heron and Reason, 2001: 7).

The principle of researching-with may be weak, especially where practitioners are using their own practice as the research focus because they have primary ownership of the inquiry. The working-with aspect of local practitioner research can be honored by ensuring that affected students and colleagues are aware of the project aims and activities and invited to offer their views and reflections throughout the research cycle. It is a question of working within the spirit of this framework if a strong model of co-research is difficult to achieve.

Another connected problem to that of the "researching with" principle is the assumption that recruiting people with first hand experience of an issue allows for a better understanding of it. At first glance, the idea that people experiencing an issue can see its nature better than others has immediate appeal but we could do with digging beneath this proposition. There is no clear evidence that experiencing a problem equips you with special glasses with which to see it. Indeed experience can be so near to the experiencers that they are barely aware of its shape and presence. Those who have direct experiences of an issue might well have particular insights about it but they might equally offer a victimist or triumphalist explanatory narrative. Similarly, insider practitioner knowledge needs to be explored for its limitations as well as its strengths. Researching-with has to be in the spirit of open inquiry rather than simple, researcher deference to the experiencers.

While dialogic forms of inquiry are likely to yield deeper understandings than the kind offered by Berger above, much will depend on the skills and openness of the co-inquirers to enter into critical, reflective conversations. Such inquiry also includes a need to be alert to the risk that the group dynamics encouraged by action research can produce partisan readings of experience which may slide easily into dogma. When a group of people get together with a shared interest, be it to make jam or conquer the world, they tend to converge in their thinking and opinions (Foss, 1996). Local knowledge can just be localized knowledge. Winter (1998: 66) suggests a way of avoiding this by:

> Placing data from a specific situation in a wider social context, looking
> for tensions and contradictions in the data and considering how these

contradictions may both reflect the history of the situation and may also be symptomatic of possible changes in the future.

Here Winter reminds us that research-with is not simply about the gathering and analysis of primary data relating to immediate experiences. Like any other research, action research needs to be enriched both by attention to secondary data and to the theoretical. Such attention is likely to involve an examination of relevant existing research, possible explanatory frameworks, depictions of the issue in currency and rival explanations to those suggested by local data. This contextual and theoretical aspect of co-inquiry can be addressed at the reconnaissance and planning phase (described below) so that it infuses the research cycle.

Having raised some issues about the democratic aspirations of action research, I do not wish to overdo my caution. Indeed, I agree with Heron and Reason (2000) that research-with has transformative potential.

Transformative/Informative

Picture, on the one hand, the difference between an interview conducted by an academic which is extractive of information *from* students according to the academic's questions and, on the other hand, a reflective dialogue *with* students as discussed above. The difference here captures the contrast suggested by action researchers between research which aims to be informative and research which aims to be transformative (Heron and Reason, 2001). In the former, the students assist an inquiry rather like witnesses to a police investigation. In the latter, the reflective dialogue is designed to enable shared understandings. The transformative character of action research comes from working with these shared understandings to negotiate possible solutions. These possible solutions are the action in action research. As indicated, the change and ways of researching it is *the* distinctive character of action research. Unlike experimental design, the proposed change should not be regarded as a "treatment" that is to be tested through laboratory-like conditions. The problem of the Hawthorn effect (where observed behavior becomes changed behavior) does not concern the action researcher because there is no concept of the contamination of findings. If the effects of observing their behavior changes it, then that is to be noted by the co-researchers as data on which to reflect further; it does not threaten the trustworthiness of the account.

Action Research as a Journey

Another way of illuminating the contrast between experimental design and action research is to contrast Brew's (2001) metaphors of research as "product" or "journey." In experimental design, the researcher does not expect to be changed by the experimentation and the aim of his or her testing is to produce a finding (the product) and then move on to the next research project, a little wiser about what works in his or her field, but not substantially changed as individuals. By contrast, action research fits Brew's notion of research as a "journey" in which the researchers position themselves as learners throughout the process, expecting to be transformed as well as to transform. Finally, the experimental design process is linear and prescribed whereas action research is a more open, cyclic journey.

The Generative Capacity of Action Research

Those with a scientific background often have trouble seeing action research as credible precisely because it breaks so many taken for granted rules of scientific inquiry, but it has its own rules of rigor and notions of trustworthiness. Within these rules it is important to bear in mind that most versions of action research do not aspire to discover truths. The aims of action research are to generate fresh understandings and new practices; these are likely to be seen as provisional and inviting review through a spirit of continual reflection. While there are superficial similarities between an experimental design treatment and an action research change intervention, experimental designers test a treatment with one group, comparing this with a group that does not receive the treatment (the control group). As far as possible, the experimental designer establishes the treatment within laboratory-like conditions, minimizing "contaminating" variables. Action research takes place in a naturalistic setting where there is little researcher control over the environment and an acceptance of its complexity.

Action researchers believe, following Lewin (1946), that the best way to understand a problem is to try and change it. For instance, an action research inquiry may discover that the purposeful facilitation of mixed learning groups of international and home students (a proposed solution) supports deeper learning for both sets of students. However, with the exception of a narrow "what works" version of action research, the hope is also that the many conversations, observations, interviews, etc. which centered on the formulation and implementation of the change generate further theoretical and practical understandings. Further, strong participatory and emancipatory models of action research regard the development of the co-researchers intellectual and practical capacity to be as important as the development of solutions. This is why action research has much appeal in the field of community development.

Trustworthiness

Given that action research houses a variety of methods, quantitative and qualitative, claims to the trustworthiness of the research will be associated with the methods selected to include a defence for their selection. The overall narrative must link the diverse data sources to show how they point towards the claims made. Alternatively, action research reports can be plurivocal, displaying tensions in the data and different ways of experiencing the issue and solution.

Method

The Action Research Spiral

In conventional research designs there is a linear, sequential relationship between question or hypothesis formulation, research activity, research findings and changed practices; in action research these activities are dynamically related through the research focus on a proposed solution. Many action researchers offer the image of a spiral to capture constant movement between the phases of reconnaissance, planning, acting, observing and reflecting.

There is a bewildering amount of debate about how these phases are conceptualized and described with different proponents offering competing diagrams alongside particular epistemological defences. But they all have in common a commitment to critical reflection throughout the research cycle and to researching change while it is happening. The promised cross breed version now follows. I hope it will allow the reader to develop an action research project without having to plunge unduly into this huge literature (much of which is online), though I do, of course, provide a further reading section at the end of this chapter.

Reconnaissance and Planning

Write a Statement of the Problem Begin with a statement of the idea or problem in hand in the form of a brief paragraph. Let us assume that the research project concerns international students. The statement might read something like:

> While international students report high satisfaction with their study experience, the majority register disappointment with the lack of opportunities to mix with home students. Valuable opportunities for home and international students to learn from each other do not appear to be embedded in pedagogic strategies or curriculum design in this university. This reflects a common reality in other universities.

This is a provisional statement that helps the team or the individual to delineate the issues to be researched to support the direction of the project.

Preliminary Research

The next step would be for the co-inquirers to discuss how they might do some preliminary research; how much they do will depend on resources available but it might include a review of relevant literature and case studies; or it might involve some preliminary interviews/discussions with key people. It could just involve a single interview with a known expert. This phase could also be about coming up with a provisional theoretical direction to help the inquiry. For instance, the literature on international students might offer a typology of how international students are regarded (e.g. guests, strangers, sojourners, teachers, exiles) and the action research can critically explore this typology as it researches its proposed change. This might result in the rejection or adaptation of the typology but it will give the researchers a preliminary mapping to support their own theorizing. An alternative would be to research promising theoretical perspectives as the data gathering proceeds. However this is managed, do not overlook the fact that the quality of action research depends on a sustained commitment to read so that the project reflexivity dynamically links practical and bookish activity.

Formulating the Research Question

Once the researchers feel they have a preliminary handle on the issue, they then need to formulate the action research question. Quite simply, this centers on the proposed change. The first statement can now be supplemented to read, for instance:

> If we develop a friendship scheme among first year undergraduates, how might this increase formal and informal learning between home and international students?

Note, by the way that this is not a hypothesis which obliges the researchers to look for causal links. The "how might" in the question invites the more realistic aspiration to search for possible links between a friendship scheme and learning. Action researchers tend to side with the increasing number of researchers who acknowledge that there are often too many confounding variables for reliable assertions about causality to be made in the human sciences. This does not mean that action researchers cannot point to possible associations between the change intervention and the desired outcome.

Planning the Change

Having formulated the change intervention (in my example the friendship scheme), the next activity is to plan how the change can be implemented. I have split this into: a) organizational; b) who-with; c) risks; and d) ethics.

Organizational

In the case of a research team, one or two coordinators need to be named so that there is clear ownership of the organizational side of the research. If there are only two of you in the team, there still needs to be a clear understanding of respective responsibilities. Getting this side sorted at the beginning will allow more enjoyment and free flow of the research process. Questions to be addressed:

1. When will the change happen and what is the timescale (a timeline needs to be produced)?
2. How is available funding going to be best used?
3. If appropriate to the research design, how and when will participants be recruited? How many?
4. Can anyone lead/manage a virtual dimension to the project (for discussions, resources, key documents, etc.)?
5. Who will lead the support for the intellectual dimension (e.g. suggest readings)?
6. What is the team communication strategy?
7. How do we ensure the manageability of the project?

Do not turn the project into a hyper-busy one. It is always better to do a small scale project thoroughly than an ambitious large scale project on stretched resources.

8. Who will collect what data and where will the data be stored and shared? Having a website (which might be an existing electronic learning environment) seems sensible and to facilitate data analysis, software is strongly advised. You will need to check whether there is appropriate skills in the team. If not, who is willing to acquire and share it?
9. How will the reflective process be captured alongside the meetings? Will, for instance, participants agree to keep diaries and field notes?
10. How many meetings will there be throughout the cycle? When will they take place?

Where there is a research team, regular meetings (real or virtual) will need to be fixed throughout the cycle to ensure the "co" part of the "co-inquiry" is meaningful and the reflective spirit of action research honored. These meetings will enable the sharing of data and analysis, the solving of emerging difficulties, the pooling of ideas, identification of further research activities and so forth.

Who-With

What is the "research-with" dimension? To stay with my example, should the research team be expanded to incorporate the under-graduates to be recruited to the friendship scheme? The most manageable expansion might be to co-opt a few of the participating students onto the research team and to be clear with all the other participating students that this is an action research project in which their views, solicited or not, will be welcome at all times.

Bear in mind that there are different levels of co-inquiry and the key issue is to find ways to be dialogic rather than information extractive. For instance, you might want to give the students a regular chance to see draft research reports on which to comment but not involve them in team meetings. Partly these decisions are practical ones.

As indicated earlier, if you are largely the sole researcher because you are investigating your own practice, this question is likely to concern the extent to which other stakeholders will be alerted to the research and given opportunities to reflect with you in a light version of co-inquiry.

Risks?

What are the risks to the project? This involves the usual ones of participants dropping out, delays or problems with ethical clearance but particular to action research are the risks attached to the proposed change. Thinking about the risks associated with planning, implementing and researching the change supports the refinement of the research design.

Ethics

An ethical framework will need to be drawn up of course. This will include agreements over data ownership and author arrangements for publication. It will also be necessary to ask whether the planned change will privilege one group of students over others. Clearly, if you are teaching on a course and want to make a change that involves all your students, then there is no ethical issue. Teachers frequently adjust their teaching methods and curriculum design to grow their competence and effectiveness; action research simply extends this to a matter of systematic and public inquiry. But if you teach two cohorts on this course and want to make the change with one cohort only, this would be ethically hard to defend as well as being outside of the spirit of action research.

If the action research is being conducted by a team, then some time to discuss a shared epistemological framework will be useful. Action research teams in higher education are often cross disciplinary which means that there will be a number of cultures of inquiry represented in the team. Here is a process for handling this:

a) Every individual is given a piece of paper (or online version) on which is written the statement of the problem and the research question.

b) Each individual writes their own paragraph beneath this in answer to the question: what do you hope the research will produce in the way of intellectual and practical outcomes.

c) The statements are shared and notions such as those of trustworthiness, plausibility, theorizing and generalization are discussed.

d) If necessary, the group agree a compromise position based on respect for the different positions represented in the discussion.

Once the team begins to collect data, they can return to these statements and explore any differences in relation to specific aspects of the data. This could enrich understandings both of the data's usefulness and status and raise the intellectual capacity of the group. This point has a bearing on the write-up and questions of authorship which also need to be clarified.

Acting

The acting phase is about implementing the planned change and watching the effects of this. Be clear about the parameters of the change and the aspects of it that need to be researched. If you have launched a friendship scheme across the university, you may only have the resources to study how it is working for a small cohort of students. Or you may be able to take a layered approach in which you can collect broad sweep data (perhaps a university-wide survey) and focus in on one department. There is no privileged research tool or method and the following list is indicative of those a team might want to consider (there are separate chapters covering much of this):

- *Reflective diaries* Perhaps team members could be asked to bring their reflections on the progress of the change to every team meeting. Alternatively, the coordinator could moderate online discussion as the collective reflexivity of the group.
- *Commissioned diaries* Perhaps a selection of students can be asked to keep a diary of a particular slice of their university experience over time. This would be a valuable data source.
- *Interviews* These will need to be in the spirit of "active interviewing." See Chapter 5 on semi-structured interviews.
- *Focus group research* The key purpose of focus group research is to enable the "sharing and comparing" of ideas within a group dynamic. This aim is very well suited to action research. See Chapter 4.
- *Stories and critical incidents* This would involve asking for stories or the identification of critical incidents from those involved in the change. This could be provided both by the researchers and those involved in the change. See Chapter 6.
- *Documentary data* There are likely to be existing documents of use to the project, e.g. notes of meetings, mission statements, student evaluations, etc.
- *Observational data* There may be contexts in which the researchers want to observe what is going on although bear in mind that the observed deserve to know that they are being watched, and to share the conclusions you make from the observations. See Chapters 7 and 8.

- *Visual* This can take the form of videos, photos or drawings. These might be gathered to support explorations or they might represent data in their own right. For instance, in my example of research with international students, photos about where they live and socialize might be important additions to the data sets. See Chapter 10.
- *Quantitative* This might include surveys of international students, figures on frequency of encounters with home students, attainment patterns and so forth.

Clearly, the decision about what kind of data to collect concerns both its appropriateness for the research design as well as the practical capacity of the research team. If a team of four were researching the friendship scheme for international students, the following decisions about methods might make sense:

a) Recruit fifteen befriender/international student pairs.
b) Secure their agreement to log when they meet and what they are learning from each other.
c) Recruit two befriender/international student pairs to the research team.
d) Conduct two focus group meetings of the international students and the befrienders respectively spaced over nine months.
e) Ask international and home students at the beginning of the scheme to provide a visual map of the people they know and connect to.
f) Ask international students and home students to re-visit their visual map at the end of nine months and make any changes that have arisen as a result of the friendship scheme.
g) Circulate a provisional analysis of the data to all the students for their comments, requesting any further indicative evidence about social and/or academic impact of the scheme.

As the research progresses, some of these strategies might be abandoned, revised, supplemented or replaced. The design challenge is to keep it realistic, provisional and within the participative spirit of action research.

I should caution against the danger of gathering and analyzing data to confirm the proposed solution rather than to research it. There are a number of action research reports which offer narratives of success as opposed to analysis and this weakens the credibility of action research as a genuinely research-curious perspective (Cousin, 2000). Keeping an eye out for unintended outcomes of the change is one way of avoiding an answer-driven approach.

Observing and Reflecting

This is where the researchers focus their observations of the change and the data about it. Prompt questions might be:

a) In what ways (if at all) does the change appear to be working?
b) Can critical success factors be identified?
c) What difficulties appear to be emerging?
d) Any theoretical insights emerging?
e) Is the data pointing to unintended outcomes of the change?
f) What are the strengths and weaknesses of the data gathered?
g) How can we be sure we have not found what we are looking for?

Remember in relation to d) the researchers were going to keep an eye on their typology as a way of thinking theoretically. Did the research generate other possible categories? Were the categories unhelpful? Irrelevant? Did the research reveal that these "types" are not attributes of students but created out of the contexts established? What new concepts would better describe the conditions, orientations of home and international students?

Finally, the team (or the lone practitioner) decide:

a) What else do we need to do?
b) What revisions to the change may be necessary?

The researchers either write up the report and/or repeat the development cycle again. Perhaps the friendship scheme failed for certain groups of learners; perhaps it needs to become a more structured mentor scheme in which mentor and mentee gain some form of accreditation. The report can include what remains to be explored.

Writing Up

Nominal group technique   A preparatory process for the write-up could be that of nominal group technique to support a group synthesis of the research. Because of its emphasis on a flat research team, Cohen et al. (2007: 309) suggest the fit of nominal group technique for action research. This method involves five stages (drawn from Cohen et al., 2007: 309):

1. A member of the group acting as a coordinator provides the team with a series of statements or questions (see below).
2. Team members write down their individual responses (this could be done online).
3. The coordinator groups the responses, circulates them and invites further comment (again this could be online).
4. Team members now work in live groups to cluster and structure the responses.
5. The coordinator invites further discussion on the decisions made about cluster and structure.

Some of the helpful advice adapted from Elliott (1991: 88) below to aid the writing up of action research (particularly 7–10) could support the question and statement formulation in 1 above. Alternatively, use this as a general guide:

1. What problem was addressed and what proposed solution formed the basis of the research?
2. What were the "knowledge" aspirations of the research? (E.g. was it about exploring possibilities? Generating under- standings? Exploring what works?)
3. What data gathering methods were used? How was data analyzed?
4. What was the ethical framework?
5. What adjustments to the change were made over time?
6. What were the intended and unintended effects of the change?
7. What understandings evolved over time? What promising theoretical directions? How are they supported by the data?
8. What are the implications for practice?

9. What further research would progress solutions and/or understandings?

Finally, it has been suggested to me that the framework I have
offered might be confused with case study research. Indeed, there
are strong similarities particularly since both are approaches which
are friendly to a variety of research methods. The difference is that
case study research has a holistic concern for the case within a defined
time and space (see Chapter 9). In contrast, the heart of action research
is in the making and the watching of a change. If you recall Lewin's
(1946) advice that the best way to understand a problem is to change
it, you have captured the heart of the action research approach.

Conclusion

Action research is an excellent way of combining higher education
research and development; it also offers an inclusive method of
researching *with* rather than on others. The next chapter provides a
further slant on this approach in the form of appreciative inquiry.

Further Reading

CARN (Centre for Action Research Network) (n.d.) Available online at:
   www.uea.ac.uk/menu/acad_depts/care/carn/mission.html.
Elliot, J. (1991) *Action research for educational change*. Buckingham, UK: Open
   University Press.
McNiff, J. (1994) *Action research: Principles and practice*. London: Routledge.
Reason, P. and Bradbury, H. (2001) (Eds.) *Handbook of action research:
   Participative inquiry and practice*. London: Sage.

10
Appreciative Inquiry

Appeal

Appreciative Inquiry (AI) is a research method that is often used to underpin change management processes in businesses but it can be of equal use for the research of educational settings. Like action research, AI is solution-oriented. A key appeal of AI concerns the relative ease with which ethical clearance can be secured for its conduct because it provides a potentially unthreatening way of researching learning environments or academic cultures.

AI presents a welcome alternative to the adversarial culture of academe although expect to attract annoyance from some if you seek to introduce both the approach and its attendant language to groups of academics. Much of the AI literature is written within the genre of self-development books and as such is a little out of tune with most academic genres.

Purpose

The Scope of AI

The purpose of AI is to combine an exploration of the empirical with an exploration into the envisioning capacity of the mind. The human imagination (described below as the "dream" phase) is part of its field of inquiry.

Any initiative of any scale where you want to explore what is happening with a view to changing it or with a view to making

167

evaluative judgments will suit AI. To give some examples, I have been associated with AI research in relation to:

1. managing change in relation to virtual learning environments;
2. developing teaching and learning in a research-oriented department;
3. academic-focussed innovation among educational software developers;
4. creating an inclusive campus.

The Key Focus

A core principle of AI proposes that a problem cannot be solved within the mindset of the problem itself; transcending this mindset involves a focus on "what gives life to organizations" so as to "discover ways to sustain and enhance that life-giving potential" (Ludema et al. 2001: 189). Thus AI strives to shift our focus from our habitual problem-focussed ways of seeing to a focus on the positive in order to "heighten energy, sharpen vision and inspire action for change" (The Center For Appreciative Inquiry, 2008). If we keep asking questions about the problems in a setting, the effect is to drive our imagination into a depressing corner. If we look for what animates a setting, we ignite our thinking with possibilities rather than limitations. In this sense, our lines of inquiry are fateful (Cooperrider and Whitney, 2005).

Research or Change Management Tool?

Some regard AI as more of a change management method than a research approach. AI can be used primarily as a change management tool but it can also provide a distinctive way of gathering and analyzing data and theory building—as such, it qualifies as a research method.

The Four "D's"

AI research involves four iterative stages, namely: discovery, dream, design and destiny although an AI approach can be used for the "discovery" phase only. In this event, AI would inform an initial line of inquiry.

Theoretical Concerns

Founding Theorists

The website for the AI Commons based at the Case Western Reserve University (2008) provides an excellent range of useful references and papers. A number of the founding theorists of AI, most notably, David Cooperrider and Suresh Srivastva (1987) are from the School of Management at this American University.

AI and Action Research

AI is both an extension and a critique of action research: it extends action research as a participative, solution oriented process; it critiques action research for its tendency towards problem-centeredness. I think a number of contemporary action researchers would quibble with the AI theorization of action research as trapped in a modernist paradigm, not least because action research is a broad church that includes postmodernists. It is true, however, that AI creates a novel emphasis on what is life-giving and can be defensibly separated from action research on this basis.

Moving Towards the Light

A key principle of AI is that change and theory generation cannot be developed within a deficit model of what is not working. In this respect, Cooperrider (2001) offers his "heliotropic hypothesis" which proposes that humans gravitate naturally towards the light. This does not mean that Cooperrider (2001) has overlooked the ability of, say, fascist movements to draw in the crowds. Rather, he is proposing that the movement towards light is latent in individuals and organizations and that AI processes try to surface it through the identification and affirmation of the positive.

The AI Question

Whatever the focus of the particular research project, AI has one overarching research question, namely "what gives life here?" There will be a particular focus for this question such as "services for students with dyslexia" or "group learning" but this overarching question will inform the direction of the inquiry, initiating the "Discovery" phase of AI. If you find the vocabulary of AI a bit too New Age do not let this deter you from reading further. AI is underpinned by scientific ideas about the complex connectivity of the mind and the body and of human interaction.

Constructionism

There are two key perspectives on knowledge generation which inform AI, namely constructionist and the "new mentalism." AI researchers, following constructionism, take the view that we are as likely to *co-create* reality with others as we are to discover it. When we approach an inquiry, we do so within the limits of our existing knowledge, views, expectations, language, values and interactivity with research participants. Research is a process of negotiation with our own and other people's meanings and reality is a co-created outcome of this. We rarely have unproblematic, direct access to the truth in the field of human inquiry. AI makes a virtue out of this interpretive predicament by arguing that since we partially make realities, we might as well make good ones. Let me illustrate with Elliot's findings.

Charles Elliot (1999) asked forty-five managers to complete the sentence "My organization is" with twenty adjectives that catch the flavour of the organization. Of the responses, 72 percent were negative and only 15 percent positive (the rest were neutral). Were Elliot to proceed to analyze his data as a whole, his attention would have to go largely on the 72 percent of critical, negative or hostile words. The responses he received provide a good illustration of the constructionist notion that the questions we ask produce co-created responses. Because we are in a culture in which we have a large deficit vocabulary to describe our organizations, if a researcher asks for twenty adjectives, people will draw on this vocabulary. The researcher has got the picture she/he asked for.

For AI researchers, there is no "true experience" to be excavated by the perfect research question; each question will privilege a particular line of inquiry and hook into an existing repertoire of possible responses to it. Aware of this fact, AI researchers do not see their method as truth yielding but as generative—of fresh understandings, possibilities, practices and solutions. In sum, our direction of inquiry and the change we are likely to get are held to be inextricably linked. The very framing of our questions shapes the outcome of our research (Cooperrider, 2001; Cooperrider and Whitney, 2005). This linkage is understood within the paradigm of "mentalism."

The New Mentalism

Many neuroscientists and social scientists have been aware for some time that what we think, imagine and envision can have a powerful impact on our physical well-being. As King Canute discovered, there is no clear cause and effect relationship between willing something and its delivery but there is evidence that what we anticipate can influence what actually happens. Of particular importance are the placebo effect, the Pygmalion effect and the notions of "internal dialogue and cognitive environments" provided by the Dutch sociologist Peter Sloterdijk (1988 in Cooperrider, 2001).

The placebo effect Few doctors would dispute from their own experiences of patient recovery that mind and body are interrelated. Most are personally and professionally familiar with the effects of administering to patients a placebo, such as fake surgery or a sugar pill. In many cases, the sugar pill or the fake surgery turns out to be more effective, apparently, than a prescribed drug. On the basis of this kind of evidence, AI does not accept a mind/body dualism, arguing instead that they are dynamically linked and mutually fateful. We do not know enough about these linkages but we do know that if we think or will something to be true, it will be true in its consequences at some level (Cooperrider, 2001).

The Pygmalion effect In education studies there are well-evidenced cases about the effects of telling teachers that particular students (randomly chosen) in their classes possess academic potential. Findings have revealed that this arbitrary selection of "the able" has led to the teachers providing more emotional support, more feedback on performance and more opportunities to shine for these students (Cooperrider, 2001). If a teacher thinks there is a lid on a student's ability, she will teach to that lid; if she thinks the student has huge potential, she will choose assignments and treat the student in ways that accord with this. Learners pick up subtle (and not so subtle) messages from teachers about their potential and internalize them as their own. This dynamic creates a self-fulfilling prophecy.

Internal dialogues   Cooperrider (2001), drawing on Sloterdijk's (1988) *Critique of Cynical Reason* argues that we all have internal dialogues, collectively or individually and that these dialogues fashion, draw on and contribute to cognitive environments which determine what we regard to be important, thinkable and possible. Typically, we inhabit a number of cognitive environments (family, work, etc.) and each of them equips us with a vocabulary that either facilitates or limits our thinking. Academics are singled out as "problematists and problem-aholics" (Sloterdijk, 1988 in Cooperrider, 2001). We are, it seems, a morose breed with a limited capacity to look on the bright side of life or, as Cooperrider (2006) might say, many of us lack "appreciative intelligence."

Appreciative Intelligence

In discussing an Israeli industrialist's efforts to create prosperity and peace in the Middle East through the construction of a multicultural "capitalist kibbutz," Cooperrider (2006) suggests that some people, like this man, have a natural "appreciative intelligence." The appreciatively intelligent see a cup half full and a set of possibilities rather than obstacles. Like Martin Luther King, they are good at vision and inspiration. Applied to research, appreciative intelligence can yield powerful insights precisely because it moves beyond the confines of the past and of a problem-centered focus.

Simultaneity, Poetry, Anticipation and the Positive

Cooperrider and Whitney (2005), see the process of inquiry and change as one and the same thing. This is called the principle of simultaneity. Another principle is "poetic." By this principle, organizations are seen as books authored by its members who tell its stories within a common vocabulary. The stories and the language, of course, can be changed and with this change comes organizational change too. The anticipatory principle operates rather like the pygmalion and placebo effects described above. What we anticipate often shapes what happens. Finally the positive principle is based on the view that change for the good is mobilized by people collaborating around what excites, enthuses, pleases and inspires.

AI as a Spoonful of Sugar?

It needs to be emphasized that AI is not a way of screening out the difficult or of producing a benign "feel good" research direction. So it is not about generating a jolly Mary Poppins world. Firstly, as Cooperrider and Avital (2004: xii) argue, "Inquiry into the good or the life-generating" is "neither comfortable nor stable." Secondly, in my experience, when people tell you what is working, much can be inferred from this about what is not working.

Trustworthiness

There is no reason why AI researchers cannot adopt a reflexive stance to strengthen the trustworthiness of the research. A key issue here is that of power. When the AI researcher gathers positive data, is this dutifully given and thus shaped by compliance more than enthusiasm? I was involved in facilitating an AI workshop in which there were definite murmurs to this effect. There is an interesting challenge for reflexive AI researchers: on the one hand, they have to create a climate for heartfelt, positive testimonies but on the other hand, they have to be aware of threats to this climate. If people do raise objections to AI, it is important to clarify that its purpose is not to deny the existence of discomfort, ambivalence, conflict and uncertainty in a setting. Rather its purpose is to establish an energetic connection with the affirming side of experiences and organizations (Cooperrider and Whitney, 2005).

Finally, the fact that a different group of people proceeding through the 4 "D" phases may come up with quite different findings and different interventions presents no difficulties of validity for AI researchers. Cohort and contextual variation is an accepted given for constructionists because reality and possibility are always negotiated and situated. AI self-consciously animates that negotiation in a particular direction whereas the problem-centered focus of other research designs are felt to do so without such an awareness. As Thatchenkery (2004: 79) summarises, "method is not simply in the service of interpretation it is part of the interpretation—about where and how you look."

Method

While there are set procedures for AI, there is no reason why you cannot adapt them to suit your own circumstances. I see AI as a spirit of inquiry as much as it is a method.

Research Focus

In deciding on the appropriateness of an AI approach, you will need to think about the focus of your research. As we have seen, AI explores the past and the present in order to inform a future direction. Does this R&D thrust fit with what you want to explore? According to Cooperrider and Srivastva (1987) a setting qualifies for AI if:

- the inquiry can center on appreciating the vitality and good in an organization;
- the setting enables the findings to be applied;
- the exploration provokes new ideas and change; and
- the participants take part in the full cycle from discovery to design.

The following questions might support your thinking about whether you can fit this criteria:

a) Does my research interest concern a group of people in a particular organization or setting?
b) Do I want to explore their experiences in this setting?
c) Do I want to explore future possibilities for the setting?

d) Would it be practical to take these people (or a group of them) through the AI phases? Will I get a willing group? Is there enough time in the research cycle?

e) In the case of evaluation research—will the sponsors of the evaluation be happy with an AI approach? What would I need to do to convince them that it is more than a "feel good" perspective?

Research Participants

Like action research, AI is a form of participative inquiry, which is why you will want to think about whether you have a group that would be willing to research with you. This group can be anything from four or five to a large academic department—or indeed a combination of each. For instance, in one AI project with which I was involved, we recruited students for the "discovery" phase and then involved the academics (and the students) in the subsequent phases (Hughes, 2004).

Those who use AI as a change management tool often involve an entire organization with huge numbers of participants (Cooperrider and Whitney, 2005). I am going to assume that you will be handling more modest numbers and much smaller settings though the processes of AI can be scaled up or down.

AI Summit

One popular strategy for handling the four "D" cycles is through the AI Summit. Quite simply, this gathers together all those involved in the inquiry and over a period of four days, each of the phases is explored. This is clearly costly on time and resources but AI adherents argue that if the aim is to bring about change, it can be the best investment an organization makes. It is also possible, of course, to organize a trimmed down version of one or two days or perhaps to distribute the process over real and virtual time.

Discovery: AI Research Questions

As indicated above, the discovery phase demands a question such as "what gives life?" to a setting (Ludema et al., 2001: 193). Clearly, the question you formulate in this spirit is going to depend on the context and focus of your research. To give an example, below are questions myself and co-researchers asked for an AI into virtual learning (Cousin et al., 2005: 114):

> "Can you identify moments when things clicked? When you were enthused by something? Excited? What has engaged your interest and attention recently?"

The last question is close to the "peak moment" once advised by AI researchers. Here are some further possible questions for higher education research to give a flavor of the lines of inquiry you might pursue:

- When did you feel good about feedback on an assignment? What did you like about it?
- What were the high points of this academic year?
- What are your experiences of positive collegiality in this university?

Whatever data gathering approach you choose, you are unlikely to have much more than five or six broad questions. When trying out your questions bear in mind that it often takes longer for people to think of the positive so expect some long pauses before you get responses. Formulate follow-up prompts to nudge those who might get stuck. Bushe (1995) suggests that you aim for stories because they better facilitate the identification of insights.

Envisioning Questions

Besides formulating "peak moment" questions, you might want to give some thought to envisioning ones. See for instance, Norum (2004: 198) who came up with an excellent set of questions to put to students:

> If you were the student-in-charge of the program and could have three wishes for the program granted, what would you wish? How would your wishes be different if they were incorporated into the curriculum?

This form of imagining question rarely finds itself on an interview schedule because we are trained to dig into the empirical realm. The notion that the imagination is a research field is counter-intuitive for many researchers. Yet this question is particularly capable of exploring how learners have experienced a program because it asks them to think expansively of the future. It is also a good question from a research ethic point of view because it does not invite the student to say negative things about individual teachers.

As indicated, questions are designed to encourage participants to think about the "positive capacity" of systems, cultures or organizations (Cooperrider and Whitney, 2005). Alongside agreeing the topic focus and the questions to ask, decisions need to be made about how to collect the data; this can take the form of written stories, group, paired or individual interviews.

Writing Positive Stories

In the case of my research into virtual learning, we provided our prompt questions above and asked participants to write their stories against them. This writing took place in a workshop dedicated for this purpose. We invited colleagues to finish their stories in their own time and to send them to us against a deadline. Below are some lengthy extracts to give a flavour of the kind of responses this process elicited (Cousin et al., 2005: 114, 115).

> This story is about a middle-aged lecturer with a certain degree of optimism. She used to hate computers till the tender age of 24, but then she saw the light of the BBC micro and made it her mission in life to make sure that students could be part of her "illumination." There were major jumps over the years, from BBC Micro to Toolbook to Internet, but WebCT truly brought some "disco" lights to her teaching. She found discussion mind-blowing and students' unsolicited electronic praise of the online environment to support their studies exciting.

> Although I really enjoyed developing the Information Skills tutorials with Kathy, the real excitement began when I used the tracking facilities in [the online environment] and realized how heavily the students were using the resource. The survey results too, were astonishing, with nearly half of the students responding. And those responses were very positive.

It was then that I felt that I was doing something worthwhile and had produced something that could be expanded. The responses from my peers within other institutions have also been very positive.

With developing technology one could provide pictures. I bought some relatively cheap software from Serif to use at home and discovered how easy, though time-consuming, it was to produce animations. I was enthused again. Animations are particularly helpful in building up the steps to solving a maths or stats problem. The then new website for the Maths Support Center gave others and myself the vehicle for using these ideas. More enthusiasm. Flash animations with voice overs (for dyslexic students) on solving algebraic problems seemed another tremendous and cheering advance. Exciting.

Appreciative Stories

We found that gathering written stories from academics worked really well because it seemed to give them a particular licence to be expressive. Note the extracts above are peppered with words such as "exciting," "astonishing," "enthused again," "mind blowing," "tremendous," "cheering." We were quite sure that had we asked for "barriers to effective virtual learning" we would have ended up with the kind of data Elliot (1999) above received. Incidentally, from this standpoint, AI is also a "less is more" research approach.

Depicting the Positive

Another possible way of gathering data would be to ask participants to visually depict positive aspects, either as a basis for further discussion or as data in its own right.

Interviews—Group, Individual or Paired

The interview is the most common method by which AI researchers elicit responses. Here are some illuminating comments from Bushe (1995: 17):

> Simply talking about one's "peak" experience can easily degenerate into
> social banter and cliché-ridden interaction. The hallmark of successful

appreciative interviews seems to be that the interviewee has at least one new insight into what made it a peak experience . . . the key seems to be suspending one's own assumptions and not being content with superficial explanations given by others; to question the obvious and to do this in more of a conversational, self-disclosing kind of way.

This is what Holstein and Gubrium (1995, 1997) would refer to as an "active interview" (discussed in Chapter 5). An alternative to individual interviews would be to conduct focus group research in which you ask members to exchange stories of "peak" moments and to explore the insights they may be revealing.

Analysing Discovery Data: The Dream Phase

For the purpose of analysis, Bushe (1995) has called this phase "understanding." Firstly, the gathered stories are read by as many participants as possible and discussion centers on the appreciative principles and insights they may be revealing. Next, the participants are encouraged to craft propositional statements. These statements are a form of abstraction from the data, offering provisional theorizing about "what gives life." To illustrate from my example of virtual learning research, the statements might be:

- Acquiring C&IT capabilities that support innovative teaching is exciting.
- Academics are energised to innovate when they receive good feedback.
- Academics delight in the achievement of their students.
- Academics care about their students' learning.

Clearly these statements are highly interpretive and Bushe (1995) advises that they are circulated to those who offered the data. This becomes a phase of "amplification." Participants are asked if they seem to be fair abstractions from the stories told and whether they express aspects of the organization, project or setting? Participants can be invited to suggest further statements or to alter the ones circulated. If you are working with a "live" group to generate data, then the propositional statements can be worked up by them.

The idea of propositional statement generation is just one of many ways in which you can reduce the discovery data to manageable form.

For instance, in Hughes' (2004) research, the students summarized their findings on very creative and thought provoking posters.

In summary, the dream phase explores the data for its "positive core" and finds ways of envisioning a future from this core.

Design

In the design phase the researchers explore propositional statements, provocative, visionary or possibility propositions (Cooperrider and Whitney, 2005), such as:

- Every learner will be given a laptop so that technology can support them to achieve to their fullest potential.
- This university will graduate highly capable, employable and socially responsible learners by allowing real research opportunities throughout the undergraduate years.

Alternatively, "what if" questions can invite implementation ideas:

- What if we teachers are assigned student mentors to help them to explore e-learning?

Generating statements of this kind is best facilitated within a workshop or summit structure. Statements need to lend themselves to short, medium and long term goals. While this phase has an eye on the practical, it must be appropriately ambitious, provocative and expansive to stay within an AI spirit. The idea, write Cooperrider and Whitney (2005: 30) is to design "appreciative organizations" from the discovery and dream phase data and interpretations.

There are similarities with the Design phase of AI and the "action" phase of action research where a change is formulated.

Destiny

Finally, the last stage of appreciative inquiry is that of destiny which concerns "strengthening the affirmative capability of the whole system" (Cooperrider and Whitney, 2005: 16). This is the phase in which the dream and the design that flowed from it are realized. Organizational changes are made, people assume different roles and responsibilities and new energetic connections are made. A report is written.

Conclusion

I can imagine some readers might still have trouble seeing the iterations described above as research moves. If you do have this concern, bear in mind: a) the constructivist principle of co-creation; b) that the imagination is part of the research field; c) inquiry and change are held to be dynamically linked; and d) the limits of problem-centered research.

Further Reading

The Center for Appreciative Inquiry (n.d.) Available online at: www.centerfor appreciativeinquiry.net/what.html.

Cooperrider, D.L. (2001) Positive image, positive action: The affirmative basis of organizing. In D.L. Cooperrider, P.F. Sorensen Jr, T.F. Yaeger, and D. Whitney (Eds.), *Appreciative inquiry: An emerging direction for organization development*. Champaign, IL: Stipes Publishing.

Cooperrider, D.L. and Sivrastva, S. (1987) Appreciative inquiry in organizational life. *Research in Organizational Change and Development*, 1, 129–169.

Cooperrider, D.L. and Whitney, D. (2005) *Appreciative inquiry: A positive revolution in change*. San Francisco, CA: Berrett-Kohler Publishers.

Elliot, C. (1999) *Locating the energy for change: An introduction to appreciative inquiry*. Winnipeg, MB, Canada: International Institute for Sustainable Development.

Hughes, C. (2004) *Linking teaching and research in a research oriented department of sociology*. Project report. Available online at: www.c-sap.bham.ac.uk/resources/project_reports/findings/ShowFinding.htm?id=13/S/03.

Ludema, J., Cooperrider, D.L., and Barrett, F. (2001) Appreciative inquiry: The power of the unconditional positive question. In P. Reason and H. Bradbury (Eds.), *Handbook of action-research: Participative inquiry and practice*. London: Sage.

11

PHENOMENOGRAPHIC APPROACHES

I have entitled this chapter "phenomenographic approaches" because increasingly, research projects are inspired by phenomenography but not faithful to its "pure" form.

Appeal

Phenomenography enables the researcher to identify the range of different ways in which people understand and experience the same thing. For instance, phenomenographic findings can expose to teachers how students attempt mastery of particular concepts in their subject. Since some of these attempts are likely to be more effective than others, the appeal of phenomenography lays in the support it can offer to curriculum design (Orgill, 2008).

Purpose

Phenomenography is interested primarily in surfacing variation of experience and understanding. It is based on the following proposition:

> whatever phenomenon or situation people encounter, we can identify a limited number of qualitatively different and logically interrelated ways in which the phenomenon or the situation is experienced or understood.
>
> (Marton, 1994: 4427)

Thus the purpose of phenomenography is to identify qualitative different experiences and understandings of a particular phenomenon and among a particular sample of the population (students, teachers, etc.). It can be used to explore a vast range of issues from variation in conceptions of death to Nobel prize winners' notions of scientific intuition (Marton, 1994). See the website "Land of Phenomenography" www.ped.gu.se/biorn/phgraph/welcome.html for an extensive annotated bibliography and other resources on phenomenographic research.

Theoretical Concerns

Constructivism

The key literature on phenomenography is provided by Marton (1981, 1994) and the seminal example of phenomenography was conducted by Marton and Saljo (1976). Readers who are not psychologists might find the theoretical literature tough going but pared to its essentials, it is a research approach with a set of straightforward processes of data collection and analysis.

Etymologically, phenomenography means appearance and description (Marton, 1994). It is underpinned by the constructivist principle that we construct meanings of phenomena from an array of social and personal influences. Nobody has unmediated access to the world, it is always shaped by our experiences and our context (phenomenographers call this a "second order reality"). In short, we may not all see the same thing in the same way. For instance, one set of students might see examinations as a set of oppressive hurdles while another group might see them as offering an important means by which they pace their learning; still others might see them as ways of securing parental approval/disapproval. Some, of course, might hold a combination of these views but their leaning towards any one of them at any given time is likely to influence their study behavior. That is why variation of perception is interesting.

The aim of phenomenography in education research is to identify the various ways in which people see and experience things in order to support learning and teaching activities (as I elaborate later, these differences are referred to as "categories of description"). As indicated, what matters is not some kind of abstract truth "out there" but what people perceive to be true since this perception has practical consequences. Thus, if we discover the different ways in which students see examinations, we can formulate ways of responding to this to raise attainment levels.

In the field of curriculum inquiry, phenomenographic research often exposes the different ways in which students might understand a particular phenomenon in a specific subject area. To give an example, from interview data conducted with thirteen students, McCracken (2002, in Trigwell, 2006:) identified the following variations in relation to how his students appeared to understand geological mapping. These variations are called "categories of description":

a) A fragmented conception, where interpretation is focussed on discrete features.
b) A topographic conception, where interpretation is focussed on integration of features with previous and persistent knowledge associated with topography.
c) A geological conception, where interpretation is focussed on integration of features with new knowledge of geological processes and time dimension.

The first concern of phenomenography is to identify this kind of variation in the data. The second concern is to inter-relate these "categories of description," often in hierarchical form, in order to capture "the dimensions of variation" they suggest. In the example above, McCracken has expressed these dimensions as "fragmented," "topographic" and "geological" with the latter offering full understandings of geological mapping and the former partial ones. This "set" of interrelated conceptions is called an "outcome space." Here is Marton's summary of the process whereby this space is achieved:

> The first criterion that can be stated is that the individual categories should each stand in clear relation to the phenomenon under investigation so that each category tells us something distinct about a particular way of experiencing the phenomenon. The second is that the categories have to stand in a logical relationship with one another, a relationship that is frequently hierarchical.
>
> (Marton, 1981: 125)

Besides gradations from partial to full as with the example above, the logical relationship might also involve an "authorized" conception (perhaps the one the teacher is hoping to teach) and "common sense" conceptions (often outdated understandings of a phenomenon) (Marton, 1981) or, as Bradbeer et al. (2004) discovered in their research into conceptions of geography, the various conceptions identified might be a series of rival ones that do not yield to hierarchical organization. Although phenomenography has the potential to identify conceptions that challenge authorized versions, the researcher needs to guard against the conservativism inherent in its classificatory drive. I will return to this point below.

Deep and Surface Learning

Phenomenography has its "roots in a set of studies of learning among university students carried out at the University of Göteborg, Sweden, in the early 1970s" (Marton, 1994). Seminal research among these studies was conducted by Marton and Saljo (1976). These authors asked a number of students to read an extract from a text book, alerting them to the fact that they would then be questioned on their understandings of the text. In analyzing the responses they received, Marton and Saljo held that while some students tried to make sense of the text, others placed emphasis on memorizing it; these seemingly opposing study strategies were described as deep and surface learning respectively.

The metaphors of deep and surface learning have had a huge impact on the way faculty developers and higher education researchers have talked about undergraduate and postgraduate learning for some thirty years; they have also inspired many follow-up phenomenographic studies, some of which strive to identify the attributes of deep and surface learning in particular subject areas.

It is important to note that Marton and Saljo never claimed that deep and surface approaches are innate attributes of students; they accepted that the same student might use both approaches at different times, depending on the task in hand. Indeed, what interests phenomenographers is how our perceptions and experiences are dynamically shaped by the behavior of others and of institutions.

This point is well expressed in Beaty et al.'s (2005: 86) research into learning orientations:

> We have established that a learning orientation provides a useful construct for understanding a student's personal context for study. It encapsulates the complex nature of a student's aims, attitudes, purposes for studying. Moreover, learning orientation is not an invariable property ascribed to a student. It describes the relationship between the individual and both the course of study, the institution and indeed the world beyond the university. It can also change and develop over time.

Thus what interests the phenomenographers here is the way in which particular orientations and dispositions to study can be encouraged or discouraged by pedagogical and institutional practices. The lesson to be taken from Marton and Saljo's (1976) study is not so much to persuade students to take a deep approach to learning but to encourage teachers to teach in ways that invite such an approach (Prosser and Trigwell, 1999).

Student Experience Surveys

One of the outcomes from Marton and Sajo's (1976) influential research has been the development of many course experience questionnaires which aim to capture whether deep or surface learning is taking place (see www.tla.ed.ac.uk/etl/publications.html#measurement for a review of available surveys).

Classification and Binary Opposites

One of the criticisms of phenomenographic research is that the variations it identifies often result in neatly hierarchized categories and/or binary opposites (such as deep and surface learning) that erase more nuanced and complex understandings (Webb, 1997). Further, those categories of description which are discredited by the researcher as erroneous or partial simply may not fit with established ways of seeing and/or his own view of the subject. In this way, an ideological source of variation can be misrepresented in phenomenography as a cognitive one. Webb (1997) argues, then, that the variation in categories of description, are often an expression of the contested positions within a subject.

Webb's critique suggests that the derived categories from phenomenographic analysis must be regarded as "heuristic devices," that is, initial tools for thinking about the issue in hand, to include a critical examination of possible questions of power lurking beneath the identified variation. For instance, where relevant, the researcher could take a close look at the paradigm shifts and key disputations in his or her subject in order to explore whether the "dimensions of variation" can be associated with any of these factors. Like any research approach, the quality of a phenomenographic study will depend on the thoughtfulness brought to the collection and analysis of the data, to include some consideration for rival explanations of this sort. Another move might be to resist placing the categories of description in a hierarchy, representing them instead in linear form to suggest a more overlapping, competing and messy interrelation (Morris, 2006).

Emotional Dimension

One of the concerns I have about phenomenography is its tendency to ignore the emotional dimension to learning because much of it is concerned with conceptual understandings. It could be, for instance, that surface learners are more scared and less confident than deep learners but the categories of description tend not to capture this dimension; they usually display a cognitive range though there is no reason why the phenomenographic method cannot include the affective elements to the variation identified.

Researcher Neutrality

A further criticism of phenomenography is that it treats the researcher as a neutral excavator of experiences from in-depth interviews. Maybe this is unfair because Marton (1994:) does see the phenomenographic interview as a dialogic and reflective event. However, when he writes that the interview "should facilitate the thematization of aspects of the subject's experience not previously thematized" you get a sense that the interviewee does indeed have a set of experiences to be excavated like pieces of wedged coal that can be loosened and seized. There is less awareness that the researcher's questions or status might affect the size and shape of the coal. In her excellent web-based guide to phenomenography, following Webb, Orgill (2008) writes:

> It is more reasonable to assume that researchers have had certain experiences and hold certain theoretical beliefs that will influence their data analysis and categorization. Webb calls for researchers to make their backgrounds and beliefs explicit, not because having these backgrounds and beliefs is bad, but rather because the readers and users of phenomenographic research need to be informed about all variables that have potentially affected the study results. My personal opinion is that such self-examination may lead to additional insights into the data and, to some extent, a more critical examination of how the researchers own beliefs have affected the research and the results of this research.

Usefully, then, Orgill proposes an updating of phenomenographic research to take account of more recent thinking about researcher positionality. Her advice amounts to a call for reflexive phenomenography that replaces the convention from some phenomenographers of "bracketing" (Morris, 2006) in which the researcher strives to set aside his own ways of seeing while gathering and analyzing the data. This ambition to "bracket" what we think in order to listen to others is very hard to put into practice. Of relevance here is Richardson's (1994) complaint that phenomenographers rarely problematize the interview as a power invested social event. Perhaps phenomenography needs to enter a new phase in which it comes to terms fully with its interpretivist nature by exposing researcher reflexivity in reports. Bowden (2008), for instance, describes his approach as "developmental phenomenology" because he sees the interview process as an occasion in which learners can advance their understandings. See also Dortins (2000) discussed below.

Trustworthiness of the Research

The empirical data gathered and analyzed in phenomenographic research supports the researcher's thinking and construction of "categories of description." What makes the results trustworthy are not claims that the empirical research has yielded a final truth about these categories. Rather, the researcher demonstrates the plausibility of the account with a critical and honest display of extracts from the data. Morris (2006) has pointed to the tendency of some phenomenographers to report their findings without supportive quotations from their interviewees. This clearly threatens the trustworthiness of the account since the reader needs to feel confident that there is a resonance between the displayed data and the claims made for it.

Another move phenomenographers sometimes make is to get other researchers to look at their data to see if they come up with similar "categories of description" or add to those already derived (e.g. Ashwin, 2006).

A final word on trustworthiness, in citing what he considers to be poor examples of phenomenographic research, Richardson (1994) argues that people should not attempt this kind of qualitative research unless they have the "proper supervision and training." I would go for a more inclusive approach, by pointing out that phenomenographic research involves craft skills that need to be developed consciously in the course of the research process, whatever their starting point—though I do understand Richardson's irritation with those who think that any kind of qualitative research comes with a licence to be sloppy.

Exhausting the Categories

A related claim to that of researcher neutrality proposes that there is a limited number of ways in which a phenomenon can be experienced and the goal of the researcher is to find these ways through in depth interviewing. Clearly our experience is always going to be shaped by historic, cultural and linguistic factors and this notion that we can capture this "limited number" may not take full account of this. There is no reason why phenomenography needs this claim and I think it is a distortion of Marton's (1994) acceptance that the variation identified will tell us what we know at a particular socio-historic moment. For me, the loose end concerns how the phenomenographer can recognize what might be radical, paradigm breaking perceptions within a hierarchical order.

Representation

Another difficulty that can be associated with phenomenographic studies is their tendency to rely on a one-dimensional subject such as "the student" or the "biologist." Although the selection of interviewees can be mindful of a range of students or biologists, phenomenographers are not interested in associating their descriptive categories with the people who provided them. Indeed, as indicated below, they treat all the interviews as a single text. Thus we can discover that some learners from an interviewed sample tend to adopt a surface approach but phenomenography does not traditionally explore whether this is linked to particular kinds of students (e.g. male or female). That said, more recent phenomenographers have incorporated an interest in the biography of the interviewees (e.g. Bradbeer et al., 2004).

Method

Research Question

Phenomenography is not hypothesis driven though it always starts with the broad speculation that variation of perception is likely to exist in relation to a given phenomenon. Nor is phenomenography like ethnography or case study research where the researcher begins with a "what is going on?" type question. Since, from the start, the phenomenographer aims to identify variation, the formulation of the research question is relatively unproblematic.

Firstly, the researchers define the "thing" they want to explore since the research question will always center on how this thing is variously experienced. Thus if you want to know something about plagiarism, your initial question will be something like "how is plagiarism perceived and experienced by students?" Once you have thought about sample size and manageability of the project, you might then refine this question to "how is plagiarism perceived by first year engineering students?" or "how is plagiarism perceived by first year international engineering students?"

Sample

The nature and size of your interview can be judged by considering what range of experience you need to cover to yield your categories of description. In the case of an inquiry into plagiarism among first year engineering students, you would need to ensure that you speak to students from a variety of study backgrounds to ensure that you capture the range of variation within this group. There is no magic sample size for phenomenographic research but at least ten interviews seems to be a sensible minimum. While you are not aiming for a representative sample, you do need to "maximize the potential variation" (Ashwin, 2006: 654) in your sample.

Sustaining a Reflexive Stance

Dortins' (2000) excellent web-available paper provides a good example of reflexive phenomenography (though she does not use this term). Entitled "Reflections on phenomenographic processes: interview, transcription and analysis," this paper is a "must read" for anyone wanting to conduct phenomenographic research. She squarely locates phenomenography in its interpretivist roots, lucidly exposing some of the epistemological and ontological issues discussed above. Her advice suggests we be mindful of the following three concerns.

- The conversational and transformational character of the interview.
- The extent to which transcripts might be cleansed up versions of the interview.
- The interpretive nature of the variation identified.

Interviews

Most phenomenographic research is based on semi-structured interviews though focus group research can be used as well as open-response survey data (see discussion of Bradbeer et al., 2004 below). In the first instance, you need to think of a way of talking about your focus in such a way as to tease out how the interviewee conceptualizes and experiences it. Here is Marton's (1994) advice:

This type of interview should not have too many questions made up in advance, and nor should there be too many details determined in advance. Most questions follow from what the subject says. The point is to establish the phenomenon as experienced and to explore its different aspects jointly and as fully as possible. The starting question may aim directly at the general phenomenon such as, for instance, when asking the subject after some general discussion, "What do you mean by learning, by the way?" Alternatively, we could ask the subject to come up with instances of the general phenomenon, asking for example, "Can you tell me about something you have learned?"

Here is an illustration of a good interview structure from Ashwin's (1996: 654) phenomenographic study at Oxford University:

> In the interviewees, academics were first asked about their experience of running tutorials generally, and then asked to describe a particular undergraduate tutorial they had given, from any preparation work they had set students through to the tutorial itself. The interviews were then structured around this description, with particular attention paid to what the tutor saw as the purpose of this particular tutorial and how they understood their role, as well as the role of student(s) within this tutorial.

Ashwin was interested in Oxford academics' perceptions of the tutorial system that is a distinctive feature of this university. Note his questions begin with a broad warm up, followed by a "grand tour" question which gets the interviewee to talk through a concrete example. This follows Marton's (1994) advice to ground the discussion:

> Most often, however, a concrete case makes up the point of departure: a text to be read, a well known situation to be discussed, or a problem to be solved. The experimenter then tries to encourage the subjects to reflect on the text, the situation or the problem, and often also on their way of dealing with it.

The only problem I can see with this advice in the case of interviewing students is the risk of staging the research event as a teacherly one. One possible route to minimize this risk would be to talk to students about an assessed task they have already carried out so that the event does not have the air of an experiment in which they are positioned as the anxious guinea pig.

Whatever strategy you adopt, you need to aim for rich, evocative, metaphoric accounts. You could experiment with graphic material to stimulate some of the discussion and to add to your data set. You could then ask subjects to visually depict their understandings and then to discuss their depictions. I have seen this done by a teacher of fashion who is currently analyzing an extraordinarily rich visual data set using this strategy.

Qualitative Data Analysis Software

It makes much sense to use software for phenomenographic analysis. It will facilitate the comparing and grouping process (as below) and allow you to assign to data chunks reflexive comments about your interpretations. It can also allow you to note the contextual factors relating to selected quotations.

Analysis of a Single Text

Although you are likely to tag each interview with some biographic detail (e.g. male, engineering student), as indicated, the idea of phenomenography is to treat all the interviews as a single text. But if you want to associate the number and nature of the identified perceptions with their provenance, there is no reason why this cannot be done within a phenomenographic approach.

Comparing and Grouping

Typically, you will go through the data three or four times to make judgments about common sets of statements and variations according to the themes you are exploring. At first, this is a bit like a card game in which you are sorting out different piles according to clubs, diamonds, etc. but each time you look at the comments you have selected for various groups, you will be adjusting, reducing and shifting them around until you are satisfied that you have fairly represented the variation. All the data chunks will be decontextualized because they will be lifted from a larger text—one challenge is to be as honest as possible about the integrity of the quote in the context of the whole interview. By the end of these comparing and grouping iterations, you will have identified your "categories of description."

Whose Categories of Description?

There is a good deal of discussion among phenomenographers about whether the categories of description can ever be more than the researchers' perceptions of the interviewees' perceptions. This is the dilemma of all interpretive methods. My advice is to not worry unduly about this as you analyze. As Dortins (2000) suggests, the variation we identify is likely to be a negotiated outcome between interviewer and interviewee. This does not diminish its worth, it simply acknowledges that no method can excavate the pure voice of the interviewee. As indicated, the categories of description are heuristic devices, they help us to advance our understanding of the phenomenon in hand, they do not have to carry the burden of being authentic in any way.

Be Parsimonious

Although you might initially identify many apparent kinds of variation, the idea is to be parsimonious, that is, to come up with around five categories of description that really seem to capture the *important* conceptions of variation being expressed by the subjects. You are not looking for variation for its own sake. You will need to make value judgments about which seem to be more telling and useful. This is a risky move because you need to make sure that you are not erasing insightful, off-center, singular perceptions in the process.

Dimensions of Variation and Outcome Space

Once you have derived your categories of description about a phenomenon, you need to place them in some kind of relationship to draw out the "dimensions of variation" they express. As indicated, these dimensions might express rival positions along a continuum or they may be organized in hierarchical form (or a combination of the two). Once you have organized your categories of description, you have provided an "outcome space." You now have the basis for discussing the implications of your analysis.

Trustworthiness

Once you have made your analysis, you can ask another researcher to check some of your interpretations. You can also show some quotes that are borderline or possible candidates for other categories to get a second opinion. Finally, you will also want to select a number of quotes to display in your report to demonstrate their fit with the categories you have generated so that the readers can make judgments about the plausibility of them.

Using Metaphors

In her phenomenographic inquiry into how academics think about research, Brew (2001: 26) began with an initial pilot to "explore, trial and then refine open-ended questions." She then interviewed 30 academics across disciplinary groupings in one Australian university. She further interviewed 27 academics across the same disciplinary groupings and across four further Australian universities. The interviews, writes Brew (2001: 26):

> sought information concerning the nature of the research participants were pursing and their experiences of research. Intervieweees were encouraged to reflect upon their research, learning and teaching.

From her analysis of the data, Brew (2001: 24, 26) identified four distinct variations and she chose the following metaphors to capture them:

Domino variation   where solving one problem can set up a chain of further problems or answers.
Trading variation   where research outputs and ideas are exchangeable commodities.
Layer variation   where discovery for the researcher is about uncovering layers of meaning hitherto concealed from view.

Journey variation   "a personal journey of discovery, possibly leading to transformation."

Brew did not suggest that researchers hold these positions neatly or discretely but by identifying these variations in her data, she was able to explore her understanding of the nature of research in the academic world; she was able to show how institutional and disciplinary cultures are in a dynamic with how we understand the research enterprise. Brew's phenomenographic research enabled her to present a case for researchers to be more self-aware about what the purpose, value and underlying theories to their activities are. Her discussion is well supported by ample quotations from the academics she interviewed.

Using Open-Response Comments from Surveys

Bradbeer et al. decided to conduct phenomenographic research on the back of a large international study of new geography undergraduates' learning styles. A questionnaire, which included a short, open-response section, was completed by 932 undergraduates in the UK, US, Australia and New Zealand. From this questionnaire, they were able to gather 153 open ended responses for phenomenographic analysis into the students' conceptions of geography and of learning and teaching. The authors do stress that they are using a phenomenographic approach rather than a pure version of it. They differed from the pure model in two key respects: first they used relatively short data sets gathered from a broader survey; second, they decided to correlate gender with the conceptions disclosed in order to explore whether variation could be associated with this.

Here is the author's description of the data analysis process in relation to conceptions of geography (Bradbeer et al., 2004: 20):

All the responses . . . were read through fairly quickly at a single sitting. They were then read a second and third time, again relatively quickly. After the third reading some tentative preliminary categories were sketched out and these were then used in a fourth and more methodical reading. Here the categories were tested against the responses and the responses themselves provisionally allocated to the categories.

The researchers then assigned descriptions to five ways in which the students appeared to understand geography, and showed them with a small sample of quotes to another phenomenographer who agreed with the analysis.

The authors found that the two most prevalent conceptions were:

- Geography is the study of the world divided into Human and Physical Dimensions.
- Geography is the study of people–environment interactions.

They also found that these two conceptions were not held by one sex more than the other. This finding allowed geographers to think about the conceptions of their subject new undergraduates bring to their studies. The analysis of the students' conceptions of learning and teaching largely confirmed the findings of Marton and Saljo (1976) discussed above although by paying attention to gender, they did discover that females were "far more likely than males to see teaching as helping learning rather than as information supply and transfer" (Bradbeer et al., 2004: 13).

Conclusion

I hope the two examples above demonstrate the purpose and usefulness of the phonemenographic approach. This approach has occupied much of the higher education research terrain especially since the 1970s. The next chapter explores recent developments in higher education curriculum inquiry in the form of threshold concept research.

Further Reading

Ashwin, P. (2006) Variation in academics' accounts of tutorials. *Studies in Higher Education*, 31(6), December, 651–665.

Beaty, L., Gibbs, G., and Morgan, A. (2005) Learning orientations and study contracts. In F. Marton, D. Hounsell, and N. Entwistle (Eds.), *The experience of learning: Implications for teaching and studying in higher education* (pp. 72–86), 3rd ed. (Internet). Edinburgh, UK: University of Edinburgh, Centre for Teaching, Learning and Assessment.

Dortins, E. (2002) *Reflections on phenomenographic process: Interview, transcription and analysis* (pp. 207–213). HERDSA Conference Proceedings.

Marton, F. (1981) Phenomenography—Describing conceptions of the world around us. *Instructional Science*, 10, 177–200.

Marton, F. (1994) Phenomenography. In T. Husén and T.N. Postlethwaite (Eds.), *The international encyclopedia of education*, 2nd ed. (Vol. 8, pp. 4424–4429). Oxford, UK: Pergamon. Available online at: www.minds. may.ie/~dez/phenom.html.

Marton, F. and Saljo, R. (1976) On qualitative differences in learning I. Outcome and process. *British Journal of Educational Psychology*, 46, 4–11.

Morris, J. (2006) The implications of either "discovering" or "constructing" categories of description in phenomenographic analysis. Proceedings from Challenging the Orthodoxies Conference, Middlesex University, December. Available online at: www.middlesex.ac.uk/aboutus/fpr/clqe/docs/Jennymorris.pdf.

Orgill, M. (2007). Phenomenography. In G.M. Bodner and M. Orgill (Eds.), *Theoretical frameworks for research in chemistry/science education* (pp. 132–151). Upper Saddle River, NJ: Pearson Education Publishing.

Orgill, M. (2008) Phenomenography. Available online at: www.minds.may. ie/~dez/phenom.html.

Richardson, J. (1994) Using questionnaires to evaluate student learning: Some health warnings. In G. Gibbs (Ed.), *Improving student learning—Theory and practice*. Oxford, UK: Oxford Centre for Staff Development.

Trigwell, K. (2006) Phenomenography: An approach to research into geography education. *Journal of Geography in Higher Education*, 30(2), 367–372.

12

Transactional Curriculum Inquiry

Researching Threshold Concepts

Appeal

A focus on threshold concepts enables academics to explore what is fundamental to a grasp of the subject they teach. They are the "jewels in the curriculum" (Land et al., 2005). A major appeal of threshold concept research lays in its focus on difficulties of mastery in the subject. Typically, it is a form of research which requires a partnership between subject specialists, educational researchers and learners. A particular appeal of threshold concept research is that it treats curriculum inquiry and curriculum design as contemporaneously feeding into each other rather than as sequential activities.

The Purpose

Broadly, the purpose of threshold concept research is to explore difficulties in the learning and teaching of subjects to support the curriculum design process. Threshold concept research begins with the working assumption that any subject will have concepts which learners will find difficult be they conceptual, emotional, psychomotor or any combination of these. This research approach does not involve a specific method of inquiry with established techniques and procedures. Rather it offers an analytical framework for bringing into view conceptual and/or affective difficulties in the disciplines. This framework can apply to any disciplinary or subject area. As we shall see, it is a theory of difficulty that proposes that mastery of a threshold concept is likely to involve both cognitive and identity shifts in the learner.

Theoretical Concerns

Threshold Concepts

The theory of threshold concepts came out of a UK wide research project which explored effective learning environments for under-graduates. This project investigated this question in five subject strands (see www.tla.ed.ac.uk/etl). The idea of threshold concepts came from Erik Meyer and Ray Land who were assigned to the economics strand. Since Meyer and Land's work, interest in threshold concept research has grown enormously with scholars from a very wide range of disciplines and many countries contributing to the field (Meyer and Land (2006, 2008).

As they investigated the teaching and learning of economics, it became clear to Meyer and Land (2003, 2005, 2006), that certain concepts were held by economists to be central to the mastery of their subject. These concepts, Meyer and Land argued, could be described as "threshold" ones because they have particular charac-teristics as described below.

Transformative

Grasping a threshold concept always involves an ontological as well as a conceptual shift. Reduced to its essential, this simply means that we are what we know. If I learn French, this does not simply involve an acquired skill set. My new knowledge becomes assimilated into my biography and thus my sense of self. I become a French speaker—and probably a Francophile. In the first stages of struggling with French, I do not self identify as a French speaker but, later, once certain understandings have "clicked" (e.g. the use of the reflexive), I start to think of myself as a French speaker rather than a learner of French. This is an important identity shift. The grasp of any subject, argue Meyer and Land, is likely to involve turning points that both deepen our understanding and bond us more closely to the subject. Thus new understandings are assimilated into our biography, becoming part of who we are, how we see and how we feel. This is an important first principle for threshold concept inquiry and it is hooked into the theorization of liminal states discussed later.

Irreversible

A threshold concept is often irreversible; once understood the learner is unlikely to forget it (this does not exclude subsequent modification or rejection of the concept for a more refined or rival understanding). One of the difficulties teachers have is that of retracing the journey back to their own days of "innocence," when understandings of threshold concepts escaped them in the early stages of their own learning. Inquiries into threshold concepts encourage academics to talk to students about what they find difficult in order to gain an appreciation of novice states.

Integrative

Mastery of a threshold concept often allows the learner to make connections that were hitherto hidden from view. Such mastery helps a learner to overcome a fragmented view of his or her subject as things fall into place.

Bounded

A threshold concept is likely to be bounded in that "any conceptual space will have terminal frontiers, bordering with thresholds into new conceptual areas" (Meyer and Land, 2006: 6). Indeed, a threshold concept can be a form of disciplinary property. For this reason, threshold concepts should be regarded as having provisional explanatory capacity. They are not fixed truths about a subject. What might be a threshold concept for economics in one phase or school of the discipline might be considered to be outmoded or erroneous by another. In my experience, the very inquiry into threshold concepts creates lively debates among subject specialists about what is central to their curriculum.

Troublesome

A threshold concept is likely to involve forms of "troublesome knowledge"; David Perkins (1999) defines such knowledge as "that which appears counter-intuitive, alien (emanating from another culture or discourse), or seemingly incoherent" (in Meyer and Land, 2003: 7). Some students protect themselves from the troubling aspects of their subject by remaining within a common sense understanding and/or by defending themselves from journeying too far into the subject. Encouraging students to abandon their intuitive understandings is troublesome because it can involve an uncomfortable, emotional repositioning. Some students might find some concepts more troubling than others (I discuss this in relation to a grasp of Otherness in cultural studies in Cousin (2006a)). Meyer and Land advance the idea of liminal states to aid our understanding of the difficulties or anxieties that attend learner mastery.

Liminal States

Meyer and Land (2006: 22) suggest that learning involves the occupation of a liminal space during the process of mastery of a threshold concept. This space is likened to that which adolescents inhabit: not yet adults; not quite children. It is an unstable space in which the learner may oscillate between old and emergent understandings just as adolescents often move between adult-like and child-like responses to their transitional status. Similarly men in a state of midlife crisis in which "there is uncertainty about identity of self and purpose in life" (Meyer and Land, 2006: 22) could be said to be occupying a liminal space.

Meyer and Land (2006) borrow the idea of liminality from anthropology (van Gennep (1960) and Turner (1960)), arguing that just as "rites of passage" mark a person's movement from one status to another (e.g. from boyhood to manhood), so disciplines require learners to enter their communities. Ultimately, their passage will be from, say, a learner of history to a historian. A complicated mix of knowledge, skills and subject identity work will go into the journey towards the status of a historian.

Once learners enter a liminal space (just think of your first lessons in French, golf, physics, etc.), they are engaged with the project of mastery. Some learners hover at the edges in a state of pre-liminality in which understandings are at best vague. Some will fake understandings (mimicry); some will frequently get "stuck" and most will oscillate between grasping a concept and then losing that grasp. Now you see it, now you don't. The recursive movements that precede mastery (which is not always achieved, of course), are expressive of dynamically related cognitive and identity shifts.

Readers of a literary bent might be drawn to the affinities Orsini-Jones (2008) makes between the concept of liminality and the allegorical journey represented in Dante's *La Divina Commedia* in which the narrator must navigate his way through a dark forest, getting lost, encountering the troubling states of hell and purgatory before he reaches paradise.

Provisional Stability

It needs to be pointed out that threshold concepts are not viewed as stable, objective elements to a subject. Their identification is always going to be context sensitive and a matter of interpretation. Moreover, difficulties might not inhere in the concept as such since for one student it might be a matter of easy mastery and difficult for another. For instance, to the disdain of their fellow Europeans on the mainland, the British are very ill-schooled in formal rules of grammar. This is doubtless why Orsini-Jones (2008), discovered that British students floundered in a linguistic dark forest while Continental learners skipped to the other side with greater ease. The usefulness of the concept of threshold concepts is in the provisional stability it allows for curriculum designers to decide what needs to be attended to both at the levels of content and pedagogic strategy.

Method

I have typified the process of investigating threshold concepts as "transactional curriculum design" (Cousin, 2008) because it involves dialogue between teachers, students and educationalists, as the examples below will show.

Research Questions

Typically threshold concept research is designed to explore the following questions:

a) What do academics consider to be fundamental to a grasp of their subject?
b) What do students find difficult to grasp?
c) What curriculum design interventions can support mastery of these difficulties?

The first task of the researcher is to get subject specialists to identify likely threshold concepts; getting academics together to explore this identification has proved to be very fruitful, not least because participants have welcomed and enjoyed the opportunity to deconstruct their subject. Indeed one of the big advantages of threshold concept research is that it animates discussion and interest among academics in ways in which more generic educational issues do not. Indeed, threshold concept research is one of the few areas of educational research which give centrality to the subject specialist rather than to the educationalists. For this reason it is proving to be a very popular way of enthusing academics about their teaching to parallel their enthusiasm about their subject. Quite simply, threshold concept research helps to fuse these two areas. Focus group research among academics has proved to be a very good starting point for threshold concept inquiry

Focus Group with Academics

A likely structure for the focus group might be as follows:

1.   Start by saying that you want participants to name any concepts in their subject which they think are crucial to its mastery and which many students find difficult.

This is likely to be a very long discussion in itself so give it plenty of time. For instance, in their exploration of threshold concepts in automotive design Osmond et al. (2008), found that colleagues had initial difficulties to name a possible concept because so much of what they held to be important was "tacit knowledge." Eventually, the group agreed that "spatial awareness" was fundamental to their subject but, again, when it came to defining this, they realized that their internalized view of it depended on an array of acquired sensibilities (such as touch and feel) that were difficult to articulate, as the researchers put it:

> Using threshold concepts as a framework has enabled the research team to *open up a dialogue* with the staff in a discipline that appears, in the main, to be relatively under-theorized.

2. When you have some candidates for threshold concepts, write them up on a flipchart. You may have one only. Then you move to discuss the proposed concepts with questions such as:

- Why are they fundamental to a grasp of the subject?
- To what extent is mastery troublesome?
- What misunderstandings do students characteristically exhibit?
- Do students offer mimicked understandings rather than real mastery?
- What is the relationship between the various threshold concepts?
- How do they help to define disciplinary modes of reasoning and explanation?
- In what ways can mastery change the learner's relation to the subject? When does this mastery typically happen?
- How do we typically teach these concepts?

Before closing the discussion, confirm that participants (or some of them) are happy to pursue the curriculum design phase of the research (described below).

3. Once you have transcribed the results from this first focus group meeting, summarize the key findings in a "report and respond" structure (Stronach and McClure, 1997). This simply means leaving a space after each main finding. Send the document in hard copy and email form to the participants and ask them to check your interpretation and to add any further comments they have in the space you have provided. Specify a return date. This gives you a further valuable data set. You can also send the report to academics who did not manage to make the meeting in case they would like to respond.

Elite Interviews

If a focus group is not feasible, then you will need to conduct individual interviews with academic colleagues. These will be akin to "elite" interviews (Gillham, 2000) because your selection of an interviewee will be based on their expertise and authority in the subject. You can use the same set of questions outlined above within a semi-structured format.

Setting a Task

Davies and Mangan (2007) set up a problem in economics, distributing it to twelve academics and twenty students across three universities. The problem was designed to explore the extent to which economic concepts were brought to bear on solving it. The responses revealed to the researchers that students were still relying on common sense understandings of economic problems rather than using the concepts (e.g. opportunity cost) they had been "taught." Predictably and reassuringly, the academics did use the abstract tools of their trade. In relation to threshold concept research, this strategy for collecting data proved to be extremely helpful.

Documentary Analysis

4.   Agree key documents to be explored for the curriculum design phase; these are likely to include relevant course documentation (particularly exam questions, course descriptions and examples of student assignments—secure their permission to explore them). Fix a date to discuss the documents, agreeing to read them first in order to have provisional responses to the following kinds of question:

- What areas of knowledge and conceptual issues appear to receive the most attention?
- What is the balance between content knowledge and conceptual mastery? Is the curriculum overstuffed?
- What kinds of understandings/misunderstandings do the student assignments exhibit?

5.   At the meeting in which you discuss these and associated questions against the documents and the students' work, the aim is to come up with an analysis of the curriculum in relation to its capacity to support threshold concept mastery. Once the group has explored this in the light of the documentary analysis, they formulate teaching and learning activities that might improve mastery on a course. The inquiry now shifts to the teaching and learning moment and to explorations with students.

6.   As you explore new teaching and learning activities designed to support threshold concept mastery, you will want to gather evaluative data on the effectiveness or otherwise of the changes. Davies and Mangan (2007) distributed an end of term questionnaire in which they simply asked:

- *What did you find the most difficult? Why?*
- *Which material in this module has most helped improve your understanding of economics?*

Again, this proved to be a very fruitful source of feedback on the difficulties the students were experiencing.

Action Researching Threshold Concepts

Orsini-Jones' (2008) excellent exploration into teaching grammar to language students followed an action research orientation. Her investigation was very student-centered in contrast to some threshold concept research which starts with academics. Orsini-Jones involved two cohorts of students (in total 128 students) over a period of two academic years. First, she identifies her starting position:

> To help students with crossing threshold concepts it is necessary to devise student-centered activities that allow them to engage both in individual and collective reflection on the troublesome knowledge encountered. The overcoming of stumbling blocks will be greatly helped by the opening up of a dialogue between students and tutors and amongst students themselves.
>
> (Orsini-Jones: 216)

And:

> The distinguishing feature of this study in comparison with the majority
> of the other literature . . . is that its outcomes are based entirely upon
> interviews carried out with students—not staff—and upon students'
> "meta-reflections"—both individual and in groups—about their learning
> experience. The identification of the threshold concept has therefore
> been entirely based on the students' voices and underpinned by a student-
> centered constructionist and dialogic approach.

Combining Curriculum Inquiry with Curriculum Design

In order to establish a context for the research into threshold concept
mastery, a team of linguists and educationists designed a "grammar
project task" in which groups of students had to create web pages in
which they displayed their analysis of a sentence according to the
Hallidayan *rank scale* (Halliday 1995 in Orsini-Jones, 2008).
"Essentially," explains Orsini-Jones (2008: 217), "this involved an
analysis of the structure of sentences, clauses, phrases and words in
terms of the item immediately below each on the *rank scale* and a
taxonomy of clauses, phrases, words and morphemes." The students
had to write their reflections on the exercise after they had compiled
the web pages. This design of the students' learning produced two
clever ways of gathering data within the curriculum process.

In exploring the web pages and the reflections, Orsini-Jones and
her colleagues were able to identify where students were having
probems. Categories of morpheme, phrase and clause emerged to be
troublesome. The overarching threshold concept seemed to be that
of the *rank scale* itself. Next, the teachers explored their provisional
analysis with the students themselves.

Interviews

The data gathered in this phase was rich in a range of ways. Firstly,
the very fact that teachers were talking to students about their difficulties supported them in their learning. It gave the students a chance
to express their anxieties and to demonstrate that despite their best
efforts, some understandings were fragile. Secondly, the interviews

exposed to the teachers where the curriculum design effort needed to go. Here is an illustrative extract (Orsini-Jones, 2008: 224):

> *First student*: I understood it in class, it was when we went away and I just seemed to have completely forgotten everything that we did on it, and I think that was when I struggled because when we were sat in here, we'd obviously got help if we had questions but . . . when it came to applying it . . . I understood the lectures and everything that we did on it but couldn't actually apply it, I think that was the difficulty.
>
> *Q*: Did you feel the same as student 1?
>
> *Second student*: Yeah. I felt lost.
>
> *Q*: In lecture times as well?
>
> *Second student*: You know, I understood the concept for about lets say 10 seconds, yes yes, I got that and then suddenly, no no, I didn't get that, you know, suddenly, like this.

These students were clearly in the liminal states described by Meyer and Land above. Recalling her image of *La Divina Commedia*, Orsini-Jones (2008: 225) sums up this issue:

> Although many students admitted they were still struggling with the rank scale concept by the end of the academic year, both the assignments that they carried out and the interview data showed that some rays of light (and understanding) were filtering through the metaphorical dark forest. However the "opening" sometimes closed down again before they had fully grasped the concept.

This recursive movement between mastery and uncertainty reminds us that research into teaching and learning must address the messy world of comprehension. In particular, threshold concept research offers a challenge to linear-based learning outcomes models.

Conclusion

The overwhelming strength of threshold concepts is precisely in the opportunities for co-inquiry it presents between subject experts, students and educational researchers; I have called this transactional curriculum inquiry to capture the negotiations between these key actors in pursuit of shared understandings of difficulties and shared ways of mastering them.

Further Reading

Cousin, G. (2006b) An introduction to threshold concepts. *Planet*, 17, December. Available online at: www.gees.ac.uk/planet/p17/gc.pdf.

Meyer, J.H.F. and Land, R. (Eds.) (2006) *Overcoming barriers to student understanding: Threshold concepts and troublesome knowledge*. London: RoutledgeFalmer.

Meyer, J.H.F., Land, R., and Smith, J. (Eds.) (2008) *Threshold concepts within the disciplines*. London: RoutledgeFalmer.

13

VISUAL RESEARCH

Appeal

It would be a simplification to suggest that the old saying a "picture tells a thousand words" is at the heart of visual research though this view is part of its appeal. Visual research provides a rich way of generating understandings and of offering telling images. The visual often connects with an emotional dimension that text-based research does not reach. Visual research can be used as a method in its own right or as a complement to text- or number-based research.

Visual-based research works well with learners who prefer graphic to text-based material, including those who are not confident with their literacy or language skills.

Purpose

A Method Not An Accessory

Many visual researchers are at pains to point out that images can provide a site of explication, they do not need words to add depth or to function as an ornament to text-based reports. Images provide a form "of content that is analytically interesting in its own right" (Wagner, 2007: 47). They are, writes Stanczak (2007: 3) "inseparable components to learning about our social worlds."

Reaching the Parts that the Word Does Not Reach

In Harper's (2002: 13) view, because, as a species, we began with pictures (e.g. cave paintings) and progressed to the word and then its written form much later, "images evoke deeper elements of human consciousness than do words." For this reason, visual researchers feel that using photos and other visual forms often elicit more expressive and emotional responses than would, for example, language-based conventional interviews. Unlike language and writing, visual forms access parts of our brain which have been in gear since we appeared on the earth.

Supporting Researcher Vision

I like Allatt and Dixon's (2004) point that researcher attention to the potential of visual data helps the researcher to "see" what he or she may not have noticed. They also point out that visual data can be an invaluable source of support for the presentation of a thick description.

Theoretical Concerns

What is Visual Data?

There are two kinds of visual data—those currently in existence and those that are produced in the research setting (Grbich, 2007: 155–156). In addition to photographs and paintings, Pole (2004a: 4) suggests that visual data should include "film video, drawings and cartoons, a whole array of web-based resources, graffiti, advertisements, clothing and packaging . . . artefacts and architecture, rural landscapes and other spaces." A summary position would be "anything we experience through the visual medium."

Visual Anthropology

The first social scientific discipline to use visual sources of data was anthropology where photos or artefacts were presented both to prompt conversations and to display in research reports. There is a revived interest among ethnographers and those associated with the relatively new field of the sociology of the visual (Banks, 2001; Pole, 2004). The use of visual data in higher education studies has been slow to take off and it is more prevalent in studies of schooling where researchers have had to think about ways of eliciting views from children. Getting children to take photos, paint and draw is an obvious way to work with them. Perhaps, the relative ease with which we can secure text-based evaluations from learners in higher education has led to a neglect of the potential of visual research. There is an exception to this. Eisner (1991: 187) points out in his very interesting book *The Enlightened Eye* that there is an irony in our reluctance to use visual images except for quantitative research where there is a routine use of "histograms, trend lines, scattergrams and flow charts."

Recent proponents of visual social science are arguing with increasing success, that there is no reason for visual data to take second place to the word. Depending on the purpose and the context of the research, this source of intelligence can be a vital means by which we understand what is going on in a given setting. There is good evidence that the verbal accounts of subjects follow a different level of recall and logic to that of their visual accounts (Samuels, 2007), demonstrating that the visual should not be treated as mere illustration of the textual.

Does the Camera Lie?

Yes it does. This is an important point made by most visual researchers who rely on photographs or film as data. The tradition of "realism" in which photos and film were felt to capture the truth tends to be rejected by contemporary visual researchers who accept that they confront the same issues of researcher selectivity and representation as researchers working with text. The photographer chooses the time, the place and the subject of his or her photo, necessarily privileging some people and things and excluding others. What the photographer presents, its sequencing and so forth will need the same reflexive engagement as the presentation of interview data, as Goldstein (2007: 64) proposes:

> that we treat photographic images in the same way a scientist treats data. No experimentalist assumes that data are perfect. Indeed, all data are assumed to have a variety of types of error . . . the question then becomes not "do these data represent reality" but rather "are the deviations from reality I know to be present relevant to the question I'm asking."

Goldstein offers the following reasons why the photo cannot give a truthful account:

1. Cameras cannot replicate human vision because the latter is three dimensional and the former is one dimensional and often in monochrome; the camera does not have the capabilities of human eyes.
2. A photograph records a brief moment in time.
3. A photograph reflects the "decided moment" chosen by the photographer.
4. One shot is usually chosen over a series of shots (temporal editing).
5. Photos are often cropped to edit out part of a shot (spatial editing).
6. The photographer makes a number of choices relating to equipment, tone, color and so forth.

Goldstein (2007: 75) summarizes with the conclusion that "every photograph is manipulated." Photos need to be considered within the context in which they were taken, their purpose and anticipated viewer response.

In sum, whatever medium is used to capture data, something is both lost and gained. Visual data does not have more problems to do with trustworthiness than does text-based data, rather it has different problems. This point is quite hard to get across to researchers who are so used to the primacy of text that the visual seems to them to be at best a source of illustration rather than explanation. In unpacking the usefulness of visual data for evaluation research, Hurworth (2004: 173) writes that it "is a matter of trying to be just as rigorous as with any other data collection technique in order to produce credible and trustworthy data." Generally this seems good advice though in taking it avoid making visual data act like textual data.

Reflexivity

Issues of researcher reflexivity equally apply to visual research. Keep a research diary and, if appropriate, write/draw field notes as you conduct your research. Goldstein's (2007: 29) comments are those of an expert photographer but helpful to this end:

> When looking at an image, first and foremost, I note my emotional response: disgust, envy, heat, sensuality—my first eye–brain impressions. I then catalog as best I can the choices made by the photographer, technical and aesthetic, before, during and after image acquisition, and I ask myself how these contribute to my response . . . what I will never do is ask, is this photo real? I know it is not.

Methods

Research Questions

Visual data can respond to broad qualitative "what is going on" type questions; it can support identity related questions such as "how do young male students see themselves?" and thematic explorations such as "How do learners experience assessment/independent study/a research-led university." Visual methods can support "looking back" (what do these pictures say about your past experience) and "looking forward" (how would you change these images?)

Here are some of the different ways in which visual data can be mined and used to support an exploration of these kinds of questions.

Reading Archive Data

Films, paintings, photos and advertisements already in existence are brought together for the generation of understandings based on the messages they may convey. For instance, a "class of 77" graduation photo, a gown and mortar board, a picture of a lectern and of a library and other symbols from a university setting might be assembled together to show ritual continuities in academic life. Visual imagery of a traditional, Ivy League university could be contrasted with that of a community college to point up—wordlessly—some of the differences.

Unavoidably, there would be a degree of researcher manipulation in the assembling of these images since he or she would be in control of their selection and sequencing. In her very helpful guide to visual data analysis Weber (2007: 2) has the following to say:

The whole is often greater than or different from the sum of the parts. A dialectical approach involves interpreting a collection of images as a whole AND analyzing each individual image, scene or metaphor. This involves examining each image closely, going back and forth from the whole to the parts, again and again. Looking at the details of one image will give you some things to look for in other images or in the whole. There may be a pattern that emerges only when you look carefully at the whole collection. How do the images "speak" to each other or relate to each other? How are they the same and how do they differ? What collective messages are they conveying?

Although the assembling of archive data can be left to "speak for itself," there is no reason why the researcher cannot add a reflexive narrative to the display of the visual. Another move might be to invite teachers and students to comment on the "collective messages" the display may be conveying to them.

The Visual as Prompt

Subjects are shown photos and other visual records or artefacts to prompt discussion about a given theme. This is a researcher controlled move in which the visual is chosen to direct the interview discussion. You might show students a photo of a degree ceremony to get them thinking about what they hope to get out of a university education. The purpose of the photo in this example would be to support future-oriented thinking from the student.

A photo of students with heads bowed, regimented desks, a watchful invigilator and signs of "quiet please" on the walls might be used to prompt discussions about aspects of assessment. Such a photo might connect the student to feelings about examinations in ways that carefully crafted questions may not tap. Using visual data as prompts is one move in an interview that will be primarily speech based. The visual will support some of the process but not the entire event. In the next method, the visual dominates the research activity.

Photo Elicitation

This is perhaps the fastest growing research method in higher education. One form of photo elicitation is where images are made or chosen from an existing archive for their depiction of "collective or institutional pasts" Harper (2002: 13). These are likely to represent salient moments in a person's and/or a community's life (e.g. wedding, graduation day). Research subjects would be asked to comment on the meaningfulness to them of these images. As already discussed, the researcher would need to defend his or her particular depiction.

Auto-Driven Photo Elicitation

Another form of photo elicitation is called "auto-driven." This is where subjects are given cameras and a theme about which to take photos which they then present and discuss with the researcher. Harrington and Lindy (1998) call this research move "reflexive photography." In their study they gave ten students disposable cameras and asked them to shoot impressions of the university. Follow-up focus group research enabled them to explore with the students their selection and representation of such impressions. Similarly, Barbara Zamorski (2002) gave ten students cameras and asked them to capture their impressions of learning in a research-oriented university. She also followed up with interviews to explore the meanings of the photographs with the students.

When the research subjects return with their photos, they can display them around a wall for all to see and you can conduct focus group research based on a discussion of the images. Alternatively, you can interview students individually, particularly if you want to draw out any differences of view among the students. Either way, you would want to keep questions to a minimum so that the reflections and conversations are centered on the visual.

Give some thought as to where you will place the images so that they facilitate a focus on them. It is best to ask the subjects to handle their display and sequencing. If you have too many photos, your first step may be to get the subjects to select those that they think are the most meaningful.

In presenting the benefits of auto-driven facilitation in his research with novice monks, Samuels (2007: 204) comments that:

> One of the most fruitful benefits of autodriven photo elicitation is that it led to a breaking of my own frames. For example, had I used my

own photographs to explore the place and roles of temple duties and rituals in the lives of young monastics, I would have inserted photographs of what I considered key temple duties.

The photos shot by the novice monks revealed areas of concern which were outside of Samuels' purview, suggesting to him that his own production of photos would have excluded important issues. Indeed, Samuels (2007: 221) concludes that auto-driven photo elicitation provides the "interviewees with the opportunity to present their own worlds by becoming arbiters of their own experiences and actively involved in the construction of meaning."

Key questions might be:

- Can you take me through how you handled this photo assignment?

This is a bit of a "grand tour" warm up but also designed to discover the challenges and concerns the subject may have experienced in taking on the task of choosing and shooting settings for the research.

- Tell me about this image?

This is a standard question both in teaching and in research to get subjects to offer their interpretations of visual images. You may have to follow up this question with the prompt "that is interesting, can you tell me more?" Aim for the fullest discussion possible and bear in mind that you might want the subject to return to particular images to develop their thinking. Once the subject has talked about the photographs:

- Were there other images you wish you could have shot?

The subjects may have wanted to include other images but could not find appropriate moments to shoot them.

Bear in mind that the interview process based on the photos is, like any interview, a data *making* event as much as it is a discovery event. In particular, be alert to both your and the research participant's positioning with respect to the visual data; grow the reflective aspect of the event by exploring this explicitly with questions such as:

- Do you see yourself as part of this group/setting/ritual?

This kind of question prompts thinking about how the participant has positioned him or herself with respect to the setting.

– Might there be other ways of looking at this image? Would you have seen this in your first year? Might females/males see this differently?

These are "put yourself in other shoes" questions, designed to get yourself and the partipant thinking from a different angle.

– Might there be other ways of interpreting this?

This is about exploring rival interpretations to the one offered. Final questions can be summarizing and future-oriented:

– Which image (or three) is the most important from those you captured?

Besides exploring the selection of the most important photo/s, you may need to discuss possible differences between the most aesthetically pleasing image and the most telling as the participant selects. Your penultimate question could be:

– If you could, how would you change this/these images?

Depending on manageability, this could refer to the top three selected or to all of the images. This is a very powerful, future-oriented question. Where it seems feasible, you might want to ask participants to draw their alternative vision.
  As ever the final question is a mop up:

– This has been extremely interesting and useful. Are there any further comments you would like to make about that we have not covered?

If you decide to share your provisional analysis with the participants, you can arrange deadlines by which you submit this and receive their comments.

A Photo-Elicitation Ethic

The ethical framework for photo-elicitation has to be clear. Although the photos are used to support an exploration, you may want to include some of them in the research report. Decisions about this will have to be based on a judgment about whether any of the photos are invasive of anyone's privacy and an acknowledgment of those who took the photos.

Picture Elicitation

A similar research move to photo elicitation is that of *Picture-elicitation:* this is where the subject is asked to represent an experience graphically, usually in the presence of the researcher—the results can be used to prompt further discussion or presented as the research output. As indicated in Chapter 11 on phenomenography, learners' or teachers' visual representations of an experience or issue can be analyzed for the variation they may reveal.

Video Ethnography

Pink (2004) undertook research into people working at home. Firstly she interviewed participants in the normal way, tape recording the event; then she toured the participants' houses with a video, continuing to interview but this time getting their commentaries about where they worked in the home. In order to do this kind of research you clearly have to have a reason for following the participants around. Perhaps research into learning or informal spaces on campus would be amenable to this method.

Video Diaries

Holliday (2007) used video diaries to explore the presentation of self among gay men and lesbians. Each participant was asked to produce video tapes in chosen outfits and locations. This proved to be a powerful research method for the questions Holliday was pursuing in relation to sexuality and performance. It is a method that is congruent with media presentations of the confessional, particularly through reality television. As a variant on this method, I was involved in an exploration into students' perceptions of global citizenship. A "diary room" was set up in the manner of the television program Big Brother and students were invited to offer their views, one by one, in this room. Like Holliday, we found that the medium influenced the message as students relaxed into this familiar confessional box to say their piece. Like Holliday, we were a little frustrated at having captured monologues and while the contributions of the students were extremely rich, perhaps follow-up interviews would have further enriched the exploration.

The Visual as an Evaluation Tool

Hurworth (2004) explores the use of visual data to support evaluations in instances where:

- a wide range of activities has to be documented;
- there are programs which lend themselves to visual representation (performing arts, etc.);
- the material context is important (building, etc.);
- there is a need to articulate complex processes (e.g. engineering students working on a car); and
- it can offer a rich source of triangulation.

Clearly the photos the evaluator takes have to be related to the driving questions for the evaluation but I see no reason why shots cannot be taken of what looks interesting but not of clear relevance. Relevance is not always immediately apparent. Hurworth (2004) offers advice about sampling techniques—here are a few of her suggestions:

- Return to a particular scene at set intervals and/or over a set period to take photos.
- Shadow a person or group for the day, taking photos of their activities.
- Take pictures of the same thing from different angles to show different points of view (e.g. at the front of a lecture behind the teacher and at the back behind the students).

The key is to take photos which are meaningful. If you have reflections to make as you take the photos, note these in a research diary. Was there a negative reaction to your presence? Did you feel comfortable? Did you want to take further photos but felt that it would be intrusive? Was there any selection bias for the shots?

Hurworth (2004) also suggests that an evaluator include galley proofs in a report appendix so that readers can judge the basis on which selection took place.

Cartoons as Prompts

I am involved in a European project on quality enhancement in university education. At a recent two-day event with participants from most of the European countries, a cartoonist was hired to draw his impressions of the two days. Partly, we chose this method of evaluative data gathering because of the multi-lingual nature of the group and partly because I had seen the method work well in another context. The many drawings provided by the cartoonist provided a rich data source. The depictions from the cartoons were played back to participants to show what was going on from an outsider view and to prompt discussion about this.

Analyzing Visual Data

Some data analysis software packages can accommodate photos, video images and pictures to facilitate the analysis. Weber (2007) offers helpful web-based advice on how to analyze visual data. Some of her advice resonates with the grounded theory process of constant comparison in which the researcher pays attention to individual images but also to their possible relation with others. She reminds the researcher to stay focussed on the research question or issue that informs the research and to have a set of questions to ask of the images. Her questions are informed by critical theory and cultural studies; these are some of the areas her advice reminds us to cover:

- the intended purpose of the images;
- authorship of the images;
- textual and subtextual messages;
- the researcher's emotional response;
- absences and presence;
- what seems to be foregrounded? At the margins?
- questions of power conveyed by the images.

Not all of these issues will be of relevance to your research. You will want to generate your own list to cohere with your research focus. Bear in mind that the images are not neutral and questions that explore how they are culturally embedded will support the analysis.

Grbich (2007: 256, 157) offers guidance for a number of analytical moves for visual data drawing on historical analysis, structural analysis, post-structural analysis and ethnographic content analysis. The questions she poses for the latter offer a thoughtful way of proceeding which overlaps with Weber's advice. Grbich starts with the "content and context" of the image, then moves to the links the image might have with others and with any "dominant cultural values." Finally an interpretation is attempted.

Write-Up

The presentation of visual data, particularly video data is often difficult. Many journals do not have the capacity to handle visual images. However, this is likely to get better, particularly since so many journals are now online. As Holliday (2007: 277) points out, the Internet can facilitate the publication of images (moving or not), which can be hyperlinked to text.

Conclusion

Even if you do not want to get involved in full scale visual research, there are many good reasons for bringing some visual data into your research. Credibility is growing for visual research in higher education studies as increasing numbers of researchers discover its wealth either as a method in its own right or to complement others.

Further Reading

Banks, M. (2001) *Visual methods in social research*. London: Sage.

Grbich, C. (2007) *Qualitative data analysis: An introduction* (pp. 155–169). London: Sage.

Pink, S. (2001) *Doing visual ethnography: Images, media and representation in research*. London: Sage.

Pole, C. (2004b) (Ed.) *Seeing is believing? Approaches to visual research*. Amsterdam: Elsevier.

Stanczak, G.C. (Ed.) (2007) *Visual research methods: Image society and representation*. Thousand Oaks, CA: Sage.

Weber. S. (2007) Analyzing visual qualitative data. In *Image Identity and Research Collective*. Montreal: Concordia University. Available online at http://iirc.mcgill.ca/txp/?sMethodology&c=About%20based%20 research.

14

EVALUATION RESEARCH

Appeal

Evaluation research allows worth to be assigned to programs and initiatives, adding to the case for their continuing support and/or pointing to areas that need change. Evaluation research can also support the ongoing development of programs and initiatives.

Purpose

Broadly, there are three purposes to evaluation.

Formative evaluation aims to support the development of the program or initiative being evaluated and the people associated with such activities. Typically, this involves a provisional analysis of the effectiveness of the activities to support the program or initiative and advice about how they might be improved.

Summative evaluation makes final judgments about the worth of the program or initiative. Typically this involves presenting outcomes, expected and unexpected, and highlighting particular achievements (if there are any) as well as identifying any weaknesses.

Good evaluations try to fulfil both these functions though the emphasis between the two may change through the cycle of the evaluation (more formative initially with summative towards the end). Another way of framing these two differences is to see evaluation as judgment-oriented or improvement-oriented.

A further purpose of evaluation is to generate knowledge and understandings. Some people hold the view that evaluation research is not really research. While it is true that some evaluations are poor examples of research, this is a field that attracts very serious quantitative and qualitative research attention.

A crude difference between the purpose of evaluation research and other forms of academic research is that the former answers questions set by the stakeholders (often the funders), while other forms of academic research respond to the questions set by the researchers.

Methodological Concerns

This chapter will not follow the structure of the rest of this book because the research methods used for evaluation and the attendant theoretical concerns are not specific to evaluation. Decisions about which method to use and the methodological framework will depend on the setting, the purposes, the wishes of commissioners, the aims of the initiative, expectations of audiences and so forth. I hope that a number of the chapters in this book will support the making of such decisions.

At the end of the second chapter on ethics, I present issues that are particular to the position of the evaluation researcher. In this chapter I will offer a further exploration of evaluator positionality through a set of metaphors that express some of its important dimensions, namely: judge, friend, fool, mafia, geek and detective.

Evaluation Stances

Frustrated by the dominance of psychometric models of program evaluation on both sides of the Atlantic, a group of prominent North American and British evaluators met in Cambridge in 1972 to produce a manifesto in favor of a "turn" towards more formative and qualitative evaluation. Most of these evaluators had been involved in school programs. They critiqued what they called an "agricultural-botany" model of evaluation. While the measurement of growth in crops might be correlated fairly easily with a fertilizer treatment, they pointed out that learners were not plants and educational interventions were more complex than crop spray. Here are some of the things the Cambridge Group wanted commissioners of educational evaluation to recognize:

- The complexity of educational settings and thus the difficulty of relying on attempts to objectively measure activities and outcomes.
- The need to expand the evaluator's repertoire of methods to include ethnographic and dialogic forms of inquiry.
- The need to be alert to the unexpected, atypical and un-predictable.
- The need to be responsive to the needs of a range of stake-holders.
- The need for a revision of the evaluator stance as an objective outsider in favor of more facilitative and formative evaluation practices.

Much of their discussion centered on the last point above: should evaluators jet in towards the end of a program in order to make judgments? Should they follow a program throughout its cycle? Should they remain distant from the stakeholders? Would this distance prevent them from building trust among key participants in order to explore what was happening? Should they be looking at systems rather than goals? What ethical stance should evaluators take about their findings, particularly if this can lead to damaged futures for initiatives and their supporters? Over 30 years later, these questions remain critical. The core group and newcomers met for a further six conference and some are formulating new directions in response to what they see as the erosion of public trust and its impact on evalua-tions. For notes on the first conference see MacDonald and Parlett (1973) and for the case for a refreshed manifesto, see Elliott and Kushner (2007).

In my exploration of metaphors for evaluation stances, needless to say, I cannot exhaust theoretical and ethical considerations associated with evaluation research—see Guba and Lincoln's (1989) edited book on fourth generation evaluation and Marvin Alkin's (2007) more recent edited book on evaluation roots for a compre-hensive overview and chapters authored by prominent evaluators. My aim is to introduce some of the strategies and dilemmas that might confront someone undertaking evaluation research in higher education.

1. The Judge

This metaphor conjures forth the external summative evaluator who hopes to reach an objective judgment about the effectiveness of a program or initiative. Much of what passes as curriculum evaluation in our universities is judgment-oriented often with the judgments based on end of term student feedback surveys. Some of these surveys are little more than satisfaction surveys, others are more serious attempts to capture how the students have experienced their learning. For instruments of the latter kind see www.tla.ed.ac.uk/etl/publications. html#measurement.

Audit Logic

Judgment-oriented evaluation strives to set standards of performance which can be measured to inform a value judgment about the initiative or program. O'Neil (2002) refers to this focus on measurement as an "audit logic," noting its tendency to fix on what is countable at the cost of exploring experiences which are less likely to yield to a quantitative assessment.

It is important to stress that reservations about confining an evaluation to an audit logic should not deter the evaluator from using quantitative evidence or measuring outcomes. We need *some* input/output data to see what an initiative appears to have formally achieved. An evaluation framework that does not incorporate something of a judgment orientation fails its audience and fund-holders. However, initiatives and programs do not follow a linear and predictable logic. They are likely to be messy and difficult to read at times and an evaluation framework needs to ensure that it addresses this so that precious gains or indeed troubling issues do not escape their attention.

Process and Unexpected Outcomes

Besides formal countable outcomes, evaluators need to attend to process and unexpected outcomes. I once evaluated a project which had as its target the production of 80 case studies. Towards the end of the project cycle, the director expressed his disappointment that he had yet to achieve his target. Not only was he pretty close to his target figure (he had collected 70 case studies) but he had overlooked ways in which his initiative had established a strong network of academics in his field across the UK. This second outcome was both a process and an unexpected outcome from the project that needed to be acknowledged. In fixing on countable outcomes, the director was downplaying a very significant process outcome of his project.

Stake's (1967) countenance model of evaluation offers a structure for capturing both intended, unintended and process outcomes. According to this model evaluators first depict the relevant characteristics of the setting before the initiative or program began (antecedents). This provides some kind of benchmark against which to judge developments. The next task is to specify the anticipated outcomes around which indicative data will be grouped (quantitative and qualitative). Parallel to the assessment of the extent to which anticipated outcomes are met, data will also be grouped according to its ability to point to unexpected outcomes. As Deepwell (2002: 85) explains, "the congruence between the intentional and the observational provides the basis for judging the success or otherwise of the innovation, whilst at the same time allowing for the recording of unintended outcomes." Stake (a key contributor to the Cambridge Conferences) elaborated his countenance model some time ago; more recently, as we shall see, he has been a promoter of a broad notion of "responsive evaluation."

Models of transactional evaluation (Rippey, 1973) are also capable of capturing process and unexpected outcomes because they focus on an evaluation of the systems that have to deliver programs and initiatives. Rippey (1973: 3) offers an example of how this works:

> If a school system were planning to introduce a performance contracting system, transactional evaluation would look, not at improving reading scores of students, but at changed role relationships and latent appre-hensions among those responsible for the educational services—teachers, administrators and perhaps parents . . . the subject of the evaluation is the system, not the client of the services rendered by the system.

This privileging of attention to means rather than ends allows the evaluators to uncover difficulties that can be explored with stakeholders for the effective implementation of change. The next metaphor aims to express an evaluator stance that is with the stakeholders for this kind of exploration.

2. The Friend

This is where the evaluator works with project members to define the focus, gather the data and share the provisional analysis. Appreciative inquiry, action research models and emancipatory evaluation approaches would be compatible with this stance. "Fourth Generation Evaluation" (Guba and Lincoln, 1989) expresses the sentiments of the Cambridge group in seeking a decisive break from what its proponents regarded as the grip of positivism on evaluation communities. Key principles for fourth generation evaluators are that of a negotiated approach with stakeholders; the need to address issues of power; and a constructivist view of knowledge generation. Stake's (1998) outline of "responsive evaluation" also falls within this framework.

Broadly the following characteristics can be associated with a "friend" model:

- It is interested in the issues as they present themselves rather than fixing on the declared aims or goals of an initiative or program.
- It creates ongoing conversations and thinking with those who are attached to the initiative or program.
- It resists a drive towards single explanations or depictions, accepting that there might be different views and experiences to be expressed.
- It is committed to the production of evaluation reports which offer rich descriptions and analysis so that different audiences can profit from the evaluation.
- It accepts that knowledge is constructed as much as it might be discovered.

Fourth generation evaluation is in much of the spirit of the Cambridge Conference. Also within a "friend" orientation is MacClure and Stronach's (1997) "report and response" strategy. Quite simply they return a provisional report to project or program participants. Space is left between each important theme reported so that participants can add their comments. This simple iteration in the development of the evaluation report offers an invaluable way of picking up confirming or disconfirming data and of extending the evaluators' understandings.

Another "friend" approach would be Parlett and Hamilton's (1972) model of "issues-based," illuminative evaluation which tries to grapple with complex questions from the viewpoint of different players and through the provision of illuminative accounts for reader judgments to be made.

Clearly, "friend" models of evaluation are well suited to formative and development-oriented evaluation. They carry the opposite risk to that of the judge, namely that of "going native," of over-identifying with the participants of an initiative. To reduce this risk, perhaps evaluators need to position themselves as both inside and outside an initiative: sometimes getting close to participants, sometimes creating a reflexive distance. The next metaphor nicely captures the dance between inside and outside of this nature.

3. The Fool

This is the person, like Shakespeare's fool, who asks the awkward questions, discomforts the audience, disturbs conventional ways of seeing things. The fool corrects the friend's tendency to go native; the fool's job is to make the ordinary and taken for granted "strange," to explore paradoxes, ironies, tensions, the troubling and the challenging. Above all, the fool is an acute observer. The fool stance could be associated most closely with "portrait evaluation," a model supported by some members of the Cambridge Group; this model draws heavily on observational data to draw up vivid depictions of people and settings. This form of evaluation collided with objections about its ethical dimension. Some of its portraits were considered to be too invasive, perhaps even cruel like the more biting comments of a fool. But whereas the fool enjoyed regal licence to say what he wanted, the evaluator must be responsive to commissioners and participants alike.

While intrusive forms of evaluation clearly need to be avoided, it is worth keeping in mind the image of the fool to remind evaluators to be alert to their responsibility to acutely observe and to raise difficult questions. Importantly, the fool is an independent critic up to a point; what he says is meant to provoke from a position of support. Ultimately, he is not out to "do harm" for he is committed to a resolution through fresh insight. In contrast, the next metaphor comes from a dark side of the evaluation industry.

4. The Mafia

Evaluation is a business; it has societies that seek to control quality (e.g. UK, US and European Evaluation Societies). Evaluators also have informal inner circles which can be protectionist. Evaluators can overcharge, creaming off too much money from a project. They can subordinate the needs of a project or program to their research interests. They can intimidate. They can close you down. They can even sack you. There are universities who use the student voice as a gun to point to the head of teachers. Here are some questions from a university evaluation form which I took off the Internet:

> Based on your comments above, would you recommend offering this faculty member a permanent teaching position:
>
> 1. *Yes, definitely.*
> 2. *Yes, probably.*
> 3. *Unsure.*
> 4. *Probably not.*
> 5. *Definitely not.*

Apart from the ethical issues they raise (asking students to determine the career of their teachers), satisfaction surveys like this have more in common with commercial marketing practices than with educational evaluation; they tell you what students *like* as if this neatly correlates with what is pedagogically effective.

In relation to the mafia metaphor generally, some insights from Bentz and Shapiro (1998: 163) are relevant here—they cite two psychologists, Jordan and Margaret Paul (1983) who present two contrasting human dispositions: the intent to learn and the intent to protect:

> *The intent to learn* is a genuine openness to exploration and discovery, to go beyond existing boundaries in order to find out something about the other, which may sometimes involve personal discomfort.
>
> *The intent to protect* is an intention to defend one's existing boundaries, feelings and self-definitions . . . to avoid taking in anything about the other that does not fit in with one's own pre-existing feelings, beliefs, values and ideas.

The mafia are better known for their intent to protect than their intent to learn. Pawson and Tilley's (1997) model of realistic evaluation offers a good perspective on getting evaluators to adopt an intent to learn posture in their model of theory-driven evaluation. According to this model, the evaluator brings insights and provisional under-standings from his or her expertise into the project, positioning him or herself as "teacher" to discuss these; as participants respond, they in turn "teach" the evaluator. In short the evaluator and the evaluated move between teacher and learner.

Thinking about the image of the mafia is a reminder that it is a practice that concerns ethics, fair trading, value for money, trans-parency and a learning disposition. Evaluations must resist serving their own interests first.

5. The Geek

I am prompted to offer this metaphor because many evaluations, as one North American newspaper lamented, are "too boring to read." Or take the following comment from within the industry:

> At the moment, there seems to be no evidence that evaluation, although the law of the land, contributes anything to educational practice other than headaches for the researchers, threats for the innovators, and depressing articles for journals devoted to evaluation.
>
> (Rippey, 1973: 9)

Admittedly, this comment is rather old but it remains true that many evaluation reports attract more dust than action or admiration.

Imaginative Data

Perhaps the articles of which Rippey speaks are depressing not simply because they do not appear to contribute to educational practice. Perhaps they overdo what Law (2004) has called "method talk." In particular, the evaluation literature is crowded with credibility anxiety and many references to "robust" and "rigorous" evaluation "toolkits." But evaluation method is not like a piece of cutlery that you simply select for its adaptability to the food you want to eat. Rather, evaluation is a cultural practice and as such it is as much about art as it is about science. One of its aims should be to compel rather than depress the reader. The more this point gets lost on the evaluator, the duller his or her report is likely to be.

Evaluators need to offer vivid, compelling to the mind, evidence and analysis. The evaluator needs to give rich descriptive passages, vignettes, anecdotes, photographs, drawings, metaphoric usage, etc. to offer the readers a vicarious experience of "being there." Some evaluators have become over-dependent on low quality focus group and survey research, forgetting that there are other ways of yielding evidence. At least for formative evaluations, we can be creative. The key is not to censor ourselves with an over-anxious scientific model of what we are doing. The next metaphor offers a stance that encourages intuition and what Eisner (1991) has called connoisseurship.

6. The Detective

In arguing against an over rational and classificatory approach to understanding learner experiences, Guy Claxton (1997) talks about effective detectives as being capable of working with hunches as well as with empirical evidence. Take an example from the fictitious Oxford-based Inspector Morse; he was asked by a character in one episode why he wanted to look at a particular book from a murder victim's collection, "I don't know," he said, "I just stumble around." This resonates with Bentz and Shapiro's (1997: 164) account of Kurt Wolff's (1995) existential concept of "surrender and catch"—you abandon yourself to whatever you are studying and in so doing you get to catch something of the other, the subject of your enquiry.

Claxton also suggests that detectives solve their crimes through noticing the details. Good detectives do not always need huge data sets. They note that Primrose does not thank Troy as he hands her a drink—taking this as possible evidence of intimacy between the two.

Sometimes observations are all you need; or at least all you need for certain stages of evaluation. The good detective gets quality evidence from being aware, from observation. Getting to the big picture does not always require big data. The good detective respects and uses forensic science but he or she also looks in rubbish bins, ashtrays and at the books displayed on walls.

For instance, a research project in which I was involved concerned the exploration of course settings in particular institutional and disciplinary contexts. On a visit with another colleague to one such setting, our notes include the following observation:

> A radio was playing *The Archers* in one corner of this untidy and shabby room; four desks were crammed against each wall for staff . . . The window was barred up and looked out on to the back of some terraced housing; a large computer printed poster was cellotaped to the wall:
>
> > "Consultation   The procedure by which different feelings and interests are taken into account before a decision is reached."

Compare this kind of evidence with a questionnaire that might ask:

> I am involved in management decisions: very much; sometimes; occasionally; hardly ever; never.

Lickert scales for capturing experiences like this have their uses and it may be that both survey and observation would work well together. My point is that this detail on the wall supports a rich description of the morale of the department—it may be the best and most evocative sign you have, and for formative evaluation it may well be all you need to raise the issues, establish dialogue and discuss theories.

Data Swamps

There is a lot of ready to hand data in universities that can support rich descriptions for an evaluation of teaching and learning climates:

what do students write on the desks; in the toilets? What sort of authority relations are expressed in notices on the wall, such as this I found in a UK university: *The Eating of Food is Strictly Forbidden.* What does this say (apart from the odd sentence construction) about the university ethos? Is the university a scolding adult ever watchful of the rebellious young? Depressingly, I saw "I hate Asian Babes" etched into a desk at a university with a high recruitment of ethnic minority students. This is data which suggests a line of inquiry.

Inspector Morse also offers the image of what Bentz and Shapiro (1998) call the mindful inquirer; he takes himself off to listen to classical music to mull over the case. In contrast, some evaluations are very data heavy, as if data is evidence per se and certainly as if the quality of the evaluation itself can be judged by the quantity of data it has gathered. There is always the risk that researchers are looking for trophy data rather than for what is going on.

Pawson and Tilley (1997) warn against "data-driven" evaluations. This approach to evaluation research leaves little time for contemplation either for the evaluator or perhaps more importantly for the participants who are often required to express complex experiences through an infinite number of ticks in boxes. For Bentz and Shapiro and for Claxton, we need to take from Buddhist ideals a less cluttered way of researching; we need to allow for clearings in which we can think and focus with those affected by the evaluation.

The limits of the detective model, at least the fictitious variety, is that he or she is the hunter for the truth (often with the help of a faithful sidekick). He or she keeps his views to himself till the denouement, when all is explained to anxious suspects, victims and less clever colleagues. While this model of unravelling the truth has affinities with judgment-oriented evaluation which offers a "truth" at the end of the project or program, it doesn't support formative evaluation or a less truth-centered, negotiated, theory of knowledge generation. But with the detective metaphor, we can keep the hunches, the attention to detail, respect for ready to hand observational data, the commitment to surrender, to "stumble around" and to contemplate.

Conclusion

Evaluation research should not be regarded as the second cousin of other forms of educational research. Done well, it requires ethical attention, craft skills and a connoisseur's nose (to borrow from Eisner's 1991 image).

The evaluator stances above, except the Mafia, hopefully offer directions for evaluation research to support the development of a project or program and to make judgments about its worth in imaginative and intelligent ways.

Further Reading

Alkin, M. (Ed.) (2004) *Evaluation roots: Tracing theorists' views and influences.* Thousand Oaks, CA: Sage.

Claxton, G. (1997) *Hare brain, tortoise mind.* London: Fourth Estate.

Deepwell, F. (2002) Towards capturing complexity: An interactive framework for institutional evaluation. *Technology & Society*, 5(3), 83–90.

Guba, E.G. and Lincoln, Y.S. (1989) *Fourth generation evaluation.* Thousand Oaks, CA: Sage.

MacDonald, B. and Parlett, M. (1973) Re-thinking evaluation: Notes from the Cambridge Conference. *Cambridge Journal of Education*, 3(2), 74–82.

Parlett, M. and Hamilton, D. (1972) *Evaluation as illumination.* Edinburgh, UK: Centre for Research in the Educational Sciences, University of Edinburgh.

Pawson, R. and Tilley, N. (1997) *Realistic Evaluation.* London: Sage.

Rippey, R. (1973) *The nature of transactional evaluation in studies in transactional evaluation.* Berkeley, CA: McCutchon Publishing.

Stake, R. and Pearsol, J.A. (1981) Evaluating responsively. In R.S. Brandt (Ed.), *Applied strategies for curriculum evaluation*, Alexandria, VA: ASCD.

References

Adelman, C., Jenkins, D., and Kemmis, S. (1980) Rethinking case study: Notes from the second Cambridge conference. In H. Simons, *Towards a science of the singular* (pp. 47–66). CARE, University of East Anglia Occasional Publications, No. 10.

Agar, M. (1980) *The professional stranger*. New York: Academic Press.

Agar, M. (1986) *Speaking of ethnography*. London: Sage.

Alkin, M. (Ed.) (2004) *Evaluation roots: Tracing theorists' views and influences*. Thousand Oaks, CA: Sage.

Allatt, P. and Dixon, C. (2004) On using visual data across the research process: Sights and insights from a social geography of people's independent learning in times of educational change. In C. Pole, *Seeing is believing? Approaches to visual research, studies in qualitative methodology* (Vol. 7, pp. 78–104). Amsterdam: Elsevier.

Alldred, P. and Gillies, V. (2002) Eliciting research accounts: Re/producing modernist subjects. In M. Mauthner, *Feminist ethics in qualitative research* (pp. 146–165). London: Sage.

American Educational Research Association. Ethical Standards. Available online at www.aera.net/uploadedFiles/About_AERA/Ethical_Standards/EthicalStandards.pdf (accessed June 2008).

Appreciative Inquiry Commons. Available online at: www.apreciativeinquiry.case/edu/.

Ashwin, P. (2006) Variation in academics' accounts of tutorials. *Studies in Higher Education*, 31(6), December, 651–665.

Atkinson, P. and Hammersley, M. (1994) *Ethnography and participant observation*. London: Sage.

Baker, C. (1997) Membership categorization and interview accounts. In David Silverman (Ed.), *Qualitative research: Theory, method and practice* (pp. 130–143). London: Sage.

Banks, M. (2000) *Visual methods in social research*. London: Sage.

Barbour, R. (2007) *Doing focus groups*. London: Sage.

Bassey, M. (1998) Fuzzy generalisation: An approach to building educational theory. Paper presented at the *British Educational Research Association Annual Conference*, Queen's University of Belfast, Northern Ireland, 27–30 August. Available online at: www.leeds.ac.uk/educol/documents/000000801.htm (accessed April 2008).

Bassey, M. (1999) *Case study research in educational settings*. Milton Keynes, UK: Open University Press.

Beaty, L., Gibbs, G., and Morgan, A. (2005) Learning orientations and study contracts. In F. Marton, D. Hounsell, and N. Entwistle (Eds.),

The experience of learning: Implications for teaching and studying in higher education (pp. 72–86) 3rd ed. (Internet). Edinburgh: University of Edinburgh, Centre for Teaching, Learning and Assessment. Available online at: www.tla.ed.ac.uk/resources/ExperienceOfLearning/EoL5.pdf.

Bentz, V.M. and Shapiro, J.J. (1998) *Mindful inquiry in social research.* Thousand Oaks, CA: Sage.

Berger, P. (1963) *Invitation to sociology: A humanistic perspective.* New York: Bantam Doubleday Dell.

Blumer, H. (1954) What is wrong with social theory? *American Sociological Review,* 19, 3–10.

Bowden, J.A. (2008) The nature of phenomenographic research. Availabe onlone at search.informit.com.au/elibrary/phenomenography_erin/chap_01.htm (accessed April 26, 2008).

Bradbeer, J., Healey, M., and Kneale, P. (2004) Undergraduate geographers' understandings of geography, learning and teaching: A phenomenographic study. *Journal of Geography in Higher Education,* 28(1), March, 17–34.

Brew, A. (2001) *The nature of research: Inquiry in academic contexts.* London: Routledge/Falmer.

British Educational Research Association (BERA). Available online at: www.bera.ac.uk/.

Burgess, R.G. (1988) Conversations with a purpose: The ethnographic interview in educational research. In R.G. Burgess (Ed.), *Studies in qualitative methodology: A research annual* (Vol. 1, pp. 137–155). London: JAI Press.

Bushe, B.G. (1995) Advances in appreciative inquiry as an organization. *Development Intervention Organization Development Journal,* 13(3), Fall, 14–22.

CARN (Centre for Action Research Network). Available online at: www.uea. ac.uk/menu/acad_depts/care/carn/mission.html.

Carr, W. and Kemmis, S. (1983) *Becoming critical: Education, knowledge and action research.* Falmer, UK: Falmer Press.

The Center for Appreciative Inquiry (2008). Available online at: www.center forappreciativeinquiry.net/what.html.

Cheyne, A.J. and Tarulli, D. (1999) Dialogue, difference and the "third voice" in the zone of proximal development. *Theory and Psychology,* 9, 5–28.

Clandinin, D.J. and Connelly, F.M. (2000) *Narrative inquiry: Experience and story in qualitative research,* San Francisco, CA: Jossey-Bass Publishers.

Claxton, G. (1997) *Hare brain, tortoise mind.* London: Fourth Estate.

Coffey, A. and Atkinson, P. (1996) *Making sense of qualitative data.* Thousand Oaks, CA, and London: Sage.

Cohen, L., Manion, L., and Morrison, K. (2007) *Research methods in education.* London: Routledge.

Cooperrider, D.L. (2001) Positive image, positive action: The affirmative basis of organizing. In D.L. Cooperrider, P.F. Sorensen Jr., T.F. Yaeger, and D. Whitney (Eds.), *Appreciative inquiry: An emerging direction for organization development.* Champaign, IL: Stipes Publishing.

Cooperrider, D.L. (2006) Foreword: Elevating and extending our capacity to appreciate the appreciable world. In T. Thatchenkery and C. Metzker, *Appreciative intelligence: Seeing the mighty oak in the acorn.* San Francisco, CA: Berrett-Koehler Publishers.

Cooperrider, D.L. and Avital, M. (Eds.) (2004) *Constructive discourse and human organization: Advances in appreciative inquiry* (Vol. 1). Oxford, UK: Elsevier.

Cooperrider, D.L and Srivastva, S. (1987) Appreciative inquiry in organisational life. *Research in Organizational Change and Development*, 1, 129–169.

Cooperrider, D.L. and Whitney, D. (2005) *Appreciative inquiry: A positive revolution in change.* San Francisco, CA: Berrett-Kohler Publishers.

Cousin (2000) Strengthening action—Research for educational development. *Educational Developments*, 1(3), 5–7.

Cousin, G. (2006a) Threshold concepts, troublesome knowledge and emotional capital: An exploration into learning about others. In J.H.F. Meyer and R. Land (Eds.), *Overcoming barriers to student understanding: Threshold concepts and troublesome knowledge* (pp. 134–146). London: RoutledgeFalmer.

Cousin, G. (2006b) An introduction to threshold concepts. *Planet*, 17, December. Available online at: www.gees.ac.uk/planet/p17/gc.pdf.

Cousin, G. (2007) Thinking with data. *Educational Developments*, 8(1), 1–4.

Cousin, G. (2008) Threshold concepts: Old wine in new bottles? In R. Land, J.H.F. Meyer, and J. Smith (Eds.), *Threshold concepts within the disciplines* (pp. 261–272). Rotterdam: Sense.

Cousin, G. and Deepwell, F. (1998) Virtual focus groups in the evaluation of an online learning environment. ELT Conference Proceedings (pp. 16–21), University of North London.

Cousin, G., Cousin, J., and Deepwell, F. (2005) Discovery through dialogue and appreciative inquiry: A participative evaluation framework for project development. In D. Taylor and S. Balloch (Eds.), *The politics of evaluation* (pp. 109–118). Bristol: Polity.

Davies, J.W. and Cousin, G. (2002) Engineering students' writing skills. *International Conference on Engineering*, August 18–21, Manchester.

Davies, P. and Mangan, J. (2007) Threshold concepts and the integration of understanding in economics. *Studies in Higher Education*, 32(6), 711–726.

Davies Aull, Charlotte (1999) *Reflexive ethnography: A guide to researching selves and others.* London: Routledge.

Deepwell, F. (2002) Towards capturing complexity: An Interactive Framework for Institutional Evaluation. *Technology & Society*, 5(3), 83–90.

Delamont, S. (2002) *Fieldwork in educational settings: Methods, pitfalls and perspectives*. London: Routledge.

Deleuze, G. and Guattari, F. (1987) *A thousand plateaus: Capitalism and schizophrenia*. Minneapolis, MN: University of Minnesota Press.

Denscombe, M. (2007) *The good research guide*. Maidenhead, UK: Open University Press.

Denzin, H. (1984) *The research act*. Englewood Cliffs, NJ: Prentice-Hall.

Denzin, N.K and Lincoln, Y.S. (Eds.) (2000) *Handbook of qualitative research*. Thousand Oaks, CA: Sage.

Dick, B. (n.d.) *Grounded theory: A thumbnail sketch*. Available online at: www.scu.edu.au/schools/gcm/ar/arp/grounded.html#a_gt_intro (accessed March 2008).

Dortins, E. (2000) Reflections on phenomenographic process: Interview, transcription and analysis (pp. 207–213). HERDSA conference proceedings.

Doucet, A. and Mauthner, M. (2002) Knowing responsibly: Linking ethics, research practice and epistemology. In M. Mauthner, M. Birch, J. Jessop, and T. Miller (Eds.), *Ethics in qualitative research* (pp. 123–145). London: Sage.

Duncombe, J. and Jessop, J. (2002) "Doing rapport" and the ethics of "faking friendships. In: M. Mauthner, M. Birch, J. Jessop, and T. Miller (Eds.), *Ethics in qualitative research* (pp. 107–122). London: Sage.

Dunn, B. (2002) An exploration of student experiences and ethos through narratives. ELATE Conference, Coventry University, Coventry.

Eisner, E.W. (1991) *The enlightened eye: Qualitative inquiry and the enhancement of educational practice*. Basingstoke, UK: Macmillan.

Elliot, C. (1999) *Locating the energy for change: An introduction to appreciative inquiry*. Winnipeg, MB, Canada: International Institute for Sustainable Development.

Elliot, J. (1991) *Action research for educational change*, Buckingham, UK: Open University Press.

Elliott, J. and Kushner, S. (2007) The need for a manifesto for educational program evaluation. *Cambridge Journal of Education*, 37(3), September, 321–336.

Enhancing Teaching and Learning in Undergraduate Education (n.d.), Edinburgh University. Available online at: www.tla.ed.ac.uk/etl/publications.html#measurement.

Even Start (2008) Third National Even Start Evaluation, US Department of Education. Available online at: http://ies.ed.gov/ncee/pdf/20084028.pdf.

Fine, S. (2007) Student experiences: Managing dual lives in the undergraduate years. Unpublished undergraduate research project, Sussex University, Falmer, UK.

Foley, D. (1998) On writing reflexive realist narratives. In G. Shacklock and J. Smyth (Eds.), *Being reflexive in critical educational and social research* (pp. 110–129). Falmer, UK: Falmer Press.

Fontana, A. and Frey, J.H. (2000) The interview: From structured questions to negotiated text. In N.K. Denzin and Y.S. Lincoln, *Handbook of qualitative research*. London: Sage.

Foss, S.K. (Ed.) (1996) *Rhetorical criticism: Exploration and practice*. Prospect Heights, IL: Waveland Press.

Fourali, C. (1997) Using fuzzy logic in educational measurement. *Evaluation and Research in Education*, 11(3), 129–148.

Geertz, C. (1973) Deep play: Notes on the Balinese cock fight. In C. Geertz, *The interpretation of culture* (pp. 412–453). New York: Basic Books.

Geertz, C. (1983) *Local knowledge: Further essays in interpretive anthropology*. New York: Basic Books.

Gennep, A. van (1960) *The rites of passage*. London: Routledge & Kegan Paul.

Gillham, B. (2000) *The research interview*. London: Continuum.

Glaser, B.G. and Strauss, A.L. (1967) *The discovery of grounded theory: Strategies for qualitative research*. Chicago, IL: Aldine Publishing Company.

Glucksmann, M. (1994) The work of knowledge and the knowledge of women's work. In M. Maynard and J. Purvis (Eds.), *Researching women's lives from a feminist perspective*. London: Taylor & Francis.

Goffman, E. (1959) *The presentation of self in everyday life*. Garden City, NY: Doubleday.

Goldstein, B.M. (2007) All photos lie: Images as data. In G.C. Stanczak (Ed.), *Visual research methods: Image, society and representation* (pp. 61–82). Los Angeles, CA: Sage.

Gorard, S. (2006) *Using everyday numbers effectively in research*. London: Continuum International Publishing Group.

Grbich, C. (2004) *New approaches to social research*. London: Sage.

Grbich, C. (2007) *Qualitative data analysis: An introduction*. Thousand Oaks, CA: Sage.

Greenbaum, T.L. (1998) *The handbook for focus group research*. London: Sage.

Guba, E.G. and Lincoln, Y.S. (1989) *Fourth generation evaluation*. Thousand Oaks, CA: Sage.

Haggis, T. (2004) Constructions of learning in higher education: Metaphor, epistemology and complexity. In J. Satterthwaite, E. Atkinson and W. Martin (Eds.), *The disciplining of education: New languages of power* (pp. 181–197). Stoke-on-Trent: Trentham Books.

Hamilton, D., Jenkins, D., King, C., Macdonald, B., and Parlett, M. (1977) *Beyond the numbers game: A reader in educational evaluation*. London: Macmillan.

Hammersley, M. and Atkinson, P. (1983) *Ethnography principles in practice*. London: Routledge.

Hargreaves, A. (1984) Contrastive rhetoric and extremist talk: Teachers, hegemony and the educational context. In L. Barton and S. Walker (Eds.), *Schools, teachers and teaching* (pp. 303–329). Lewes, UK: Falmer.

Harper, D. (2002) Talking about pictures: A case for photo elicitation. *Visual Studies*, 17(1): 13–26.

Harrington, F.C. and Lindy, E.I. (1998) The use of reflexive photography in the study of the freshman year experience. *Journal of College Student Retention: Research, Theory and Practice*, 1(1), 399–400.

Heritage, J. (1997) Conversation analysis and institutional talk: Analysing data. In D. Silverman (Ed.), *Qualitative research: Theory, method and practice* (pp. 161–182). London: Sage.

Heron, J. and Reason, P. (2001) The practice of cooperative inquiry: Research "with" rather than "on" people. In P. Reason and H. Bradbury (Eds.), *Handbook of action research: Participative inquiry and practice* (pp. 179–188). London: Sage.

Hine, C. (Ed.) (2005) *Virtual methods: Issues in social research on the Internet*. Oxford, UK: Berg.

Holliday, R. (2007) Performances, confessions, and identities: Using video diaries to research sexualities. In G.C. Stanczak (Ed.), *Visual research methods: Image, society and representation* (pp. 255–280). Los Angeles, CA: Sage.

Holstein, J.A. and Gubrium, J.F. (1995) *The active interview*. Thousand Oaks, CA, and London: Sage.

Holstein, J.A and Gubrium, J.F. (1997) Active interviewing. In D. Silverman (Ed.), *Qualitative research: Theory, method and practice* (pp. 113–129). London: Sage.

Holstein, J.A. and Gubrium, J.F. (Eds.) (2003) *Inside interviewing: New lenses, new concerns*. Thousand Oaks, CA, and London: Sage.

Hughes, C. (2004) *Linking teaching and research in a research oriented department of sociology*, Project report. Available online at: www.c-sap.bham.ac.uk/resources/project_reports/findings/ShowFinding.htm?id=13/S/03.

Hurworth, R. (2004) The use of visual medium for program evaluation. In C. Pole (Ed.), *Seeing is believing? Approaches to visual research, studies in qualitative methodology* (Vol. 7, pp. 163–182). Amsterdam: Elsevier.

Jarrett R.L. (1993) Focus group interviewing with low-income minority populations. In D.L. Morgan (Ed.), *Successful focus groups: Advancing the state of the art* (pp. 184–201). Newbury Park, CA: Sage.

Joinson, A.N. (2005) Internet behaviour and the design of virtual methods. In C.L. Hine (Ed.), *Virtual methods: Issues in social research on the Internet* (pp. 21–34). Oxford, UK: Berg.

Kelly, E.A. and Lesh, A.R. (Eds.) (2000) *Handbook of research design in mathematics and science education*. Hillsdale, NJ: Lawrence Erlbaum Associates.

Kouritzin, S. (2002) The "half baked" concept of "raw data" in ethnographic observation. *Canadian Journal of Education*, 27(1), 119–138.

Kuhn, T.S. (1962) *The structure of scientific revolutions*, 1st ed. Chicago, IL: Chicago University Press.

Land, R., Cousin, G., Meyer, J.H.F., and Davies, P. (2005). Threshold concepts and troublesome knowledge (3): Implications for course design and evaluation. In C. Rust (Ed.), *Improving student learning 12—Diversity and inclusivity* (pp. 53–64). Oxford, UK: Oxford Brookes University.

Law, J. (2004) *After method: Mess in social science research*. Abingdon, UK: Routledge.

Lewin, K. (1946) Action research and minority problems. *Journal of Social Issues*, 2(4), 34–46.

Lincoln, Y.S. and Guba, E.G. (1985) *Naturalistic inquiry*. Beverly Hills, CA: Sage.

Ludema, J.D., Cooperrider, D.L., and Barrett, F. (2001) Appreciative inquiry: The power of the unconditional positive question. In P. Reason and H. Bradbury, *Handbook of action-research: Participative inquiry and practice* (pp. 189–199). London: Sage.

Lurie, A. (1978) *Imaginary friends*. Markham, ON, Canada: Fitzhenry & Whiteside.

McCracken, J. (1995) Implications of phenomenographic study for instructional design: A study in geological mapping. Paper presented at the 6th EARLI conference, Nijmegen, The Netherlands, August 26–31.

MacDonald, B. and Parlett, M. (1973) Re-thinking evaluation: Notes from the Cambridge Conference. *Cambridge Journal of Education*, 3(2), 74–82.

McNiff, J. (1994) *Action research: Principles and practice*. London, Routledge.

Madriz, E. (2000) Focus groups in feminist research. In N.K. Denzin and Y.S. Lincoln (Eds.), *Handbook of qualitative research* (pp. 835–850). Thousand Oaks, CA: Sage.

Marcus, G.E. (1998) *Ethnography through thick and thin*. Princeton, NJ: Princeton University Press.

Marton, F. (1981) Phenomenography: Describing conceptions of the world around us. *Instructional Science*, 10, 177–200.

Marton, F. (1994) Phenomenography. In Torsten Husén and T. Neville Postlethwaite (Eds.), *The international encyclopedia of education* (Vol. 8, pp. 4424–4429), 2nd ed. Oxford, UK: Pergamon. Available online at www.minds.may.ie/~dez/phenom.html.

Marton, F. and Saljo, R. (1976) On qualitative differences in learning. I. Outcome and process. *British Journal of Educational Psychology*, 46, 4–11.

Massey, A. (1996, 1999) Methodological triangulation or how to get lost without being found out. In A. Massey and G. Walford (Eds.), *Explorations in methodology: Studies in educational ethnography* (Vol. 2, 183–197). Stamford, CA: JAI Press.

Masters, J. (1995) The history of action research. In I. Hughes (Ed.), *Action research electronic reader*. Sydney, Australia: The University of Sydney. Available online at: www.behs.cchs.usyd.edu.au/arow/Reader/rmasters. htm (accessed December 22, 2008).

Mauthner, M., Birch, M., Jessop, J., and Miller, T. (2002) *Ethics in qualitative research*. London: Sage.

Meyer, J.H.F. and Land, R. (Eds.) (2006) *Overcoming barriers to student understanding: Threshold concepts and troublesome knowledge*. London: RoutledgeFalmer.

Meyer, J.H.F., Land, R., and Smith, J. (Eds.) (2008) *Threshold concepts within the disciplines*. Rotterdam, The Netherlands: Sense Publishers.

Miller, J. and Bell, L. (2002) Connecting to what? Issues of access, gatekeeping and "informed" consent. In M. Mauthner, M. Birch, J. Jessop and T. Miller (Eds.), *Ethics in qualitative research*. London: Sage.

Miller, J. and Glassner, B. (1997) The "inside" and the "outside": Finding realities in interviews. In D. Silverman (Ed.), *Qualitative research: Theory, method and practice* (pp. 99–112). London: Sage.

Morgan, D.L. (1997) *Focus groups as qualitative research*. London: Sage.

Morris, J. (2006) The implications of either "discovering" or "constructing" categories of description in phenomenographic analysis. Proceedings from Challenging the Orthodoxies conference, Middlesex University, December. Available online at: www.middlesex.ac.uk/aboutus/fpr/clqe/docs/Jennymorris.pdf.

Norum, K. (2004) Ap-praise-al: An appreciative approach to program evaluation. In D.L. Cooperrider and M. Avital (Eds.), *Constructive discourse and human organization: Advances in appreciative inquiry* (Vol. 1, pp. 193–216). Oxford, UK: Elsevier.

Oakley, A. (1990) Interviewing women: A contradiction in terms. In H. Roberts (Ed.), *Doing feminist research*. London: Routledge.

Oakley, A. (1999) People's ways of knowing: Gender and methodology. In S. Hood, B. Mayall, and S. Oliver (Eds.), *Critical issues in social research: Power and prejudice* (pp. 154–170). Buckingham, UK: Open University Press.

O'Neill, O. (2002), *A question of trust* (BBC Reith Lectures). BBC Radio 4. Available online at: www.bbc.co.uk/radio4.

Orgill, M. (2007) Phenomenography. In G.M. Bodner and M. Orgill (Eds.), *Theoretical frameworks for research in chemistry/science education* (pp. 132–151). Upper Saddle River, NJ: Pearson Education Publishing.

Orgill, M. (2008) Phenomenography. Available online at: www.minds.may. ie/~dez/phenom.html.

Orsini-Jones, M. (2008) Troublesome language knowledge: Identifying threshold concepts in grammar learning. In R. Land, J.H.F. Meyer, and J. Smith (Eds.), *Threshold concepts within the disciplines* (pp. 213–226). Rotterdam: Sense.

Osmond, J., Turner, A., and Land, R. (2008) Threshold concepts and spatial awareness in transport and product design. In R. Land, J.H.F. Meyer, and J. Smith (Eds.), *Threshold concepts within the disciplines* (pp. 1–17). Rotterdam: Sense.

Parlett, M. and Hamilton, D. (1972) *Evaluation as illumination*. Edinburgh, UK: Centre for Research in the Educational Sciences, University of Edinburgh.

Paul, J. and Paul, M.P. (1983) *Do I have to give up me to be loved by you?* Minneapolis, MN: Compcare.

Pawson, R. and Tilley, N. (1997) *Realistic evaluation*. London: Sage.

Perkins, D. (1999) Troublesome knowledge: The many faces of constructivism. *Education Leadership*, 57(3), 6–11.

Pink, S. (2001) *Doing visual ethnography: Images, media and representation in research*. Thousand Oaks, CA: Sage.

Pink, S. (2004) Performance, self-representation and narrative: Interviewing with video. In C. Pole (Ed.), *Seeing is believing? Approaches to visual research, studies in qualitative methodology* (Vol. 7, pp. 61–78). Amsterdam: Elsevier.

Pole, C. (Ed.) (2004a) *Seeing is believing? Approaches to visual research, studies in qualitative methodology* (Vol. 7). Amsterdam: Elsevier.

Pole, C. (2004b) Visual research, potential and overview. In C. Pole (Ed.), *Seeing is believing? Approaches to visual research, studies in qualitative methodology*. (Vol. 7, p. 1). Amsterdam: Elsevier.

Potter, J. (1997) Discourse analysis as a way of analysing naturally occurring talk. In D. Silverman (Ed.), *Qualitative research: Theory, method and practice* (pp. 144–160). London: Sage.

Potter, J. and Weatherall, M. (1987) *Discourse and social psychology*. London: Sage.

Prosser, M. and Trigwell, K. (1999) *Understanding learning and teaching: The experience in higher education*. Buckingham, UK: Open University Press.

Reason, P. (2006) Choice and quality in action research practice. *Journal of Management Inquiry*, 15(2), 187–203.

Reason, P. and Bradbury, H. (Eds.) (2001) *Handbook of action research: Participative inquiry and practice*. London: Sage.

Reason, P. and Heron, J. (2000) A layperson's guide to co-operative inquiry. Available online at: www.bath.ac.uk/carpp/layguide.htm.

Reid, I. (1977) *Social class differences in Britain*. London: Open Books.

Reissman, C.K. (1993) *Narrative analysis*. Thousand Oaks, CA: Sage.

Reissman, C.K. (2001) Analysis of personal narratives. In J.A. Holstein and J.F. Gubrium (Eds.), *Handbook of interview research* (pp. 695–710). London: Sage.

Richardson, J. (1994) Using questionnaires to evaluate student learning: Some health warnings. In G. Gibbs (Ed.), *Improving student learning – Theory and practice*. Oxford, UK: Oxford Centre for Staff Development.

Rippey, R. (1973) *The nature of transactional evaluation in studies in transactional evaluation.* Berkeley, CA: McCutchan Publishing.

Rubin, H.J. and Rubin, I.S. (2005) *Qualitative interviewing: The art of hearing data.* Thousand Oaks, CA, and London: Sage.

Samuels, J. (2007) When words are not enough: Eliciting children's experiences of Buddhist monastic life through photographs. In G.C. Stanczak (Ed.), *Visual research methods: Image, society and representation* (pp. 197–224). Los Angeles, CA: Sage

Savin-Baden, M. (2000) *Problem-based learning in higher education: Untold stories.* Buckingham, UK: SRHE/Open University Press.

Savin-Baden, M. (2004) Achieving reflexivity: Moving researchers from analysis to interpretation in collaborative inquiry. *Journal of Social Work Practice,* 18(3), November, 1–14.

Schostack, J.F. (2002) *Understanding, designing and conducting qualitative research in education. Framing the project.* Buckingham, UK: Open University Press.

Schostack, J.F. (2006) *Interviewing and representation in qualitative research projects.* Buckingham, UK: Open University Press.

Seale, C. (1999) *The quality of qualitative research.* London: Sage.

Shank, G.D. (2002) *Qualitative research: A personal skills approach.* Columbus, OH: Merrill Prentice Hall.

Silverman, D. (1997) Introducing qualitative research. In D. Silverman (Ed.), *Qualitative research: Theory, method and practice* (pp. 1–7). London: Sage.

Silverman, D. (Ed.) (1997) *Qualitative research: Theory, method and practice.* London: Sage.

Silverman, D. (1997) Towards an aesthetics of research. In D. Silverman (Ed.), *Qualitative research: Theory, method and practice* (pp. 239–253). London: Sage.

Simons, H. (Ed.) (1980) *Towards a science of the singular* (No. 10). Norwich: CARE, University of East Anglia Occasional Publications.

Simons, H. (1996) The paradox of case study. *Cambridge Journal of Education,* 26(2), June, 225–240.

Singh, G. (2007) Social research and "race": Developing a critical paradigm. Unpublished paper, *Gaps in degree attainment,* expert panel convened by the Higher Education Academy, York and the Equality Challenge Unit, London, 3–4 September, pp. 1–16.

Sloterdijk, P. (1988) *Critique of cynical reason.* Minneapolis, MN: University of Minnesota Press.

Spradley, J.P. (1979) *The ethnographic interview.* Fort Worth, TX, and London: Harcourt Brace Jovanovich.

Stake, R.E. (1995) *The art of case study research.* Thousand Oaks, CA: Sage.

Stake, R.E. (2000) Case studies. In N.K. Denzin and Y.S. Lincoln (Eds.), *Handbook of qualitative research.* London: Sage.

Stanczak, G.C. (2007) *Visual research methods: Image, society and representation*. Los Angeles, CA: Sage.

Stenhouse, L. (1975) *An introduction to curriculum research and development*. London: Heinemann.

Stewart, D.W. and Shamdasani, P.N. (1990) *Focus groups: Theory and practice*. London: Sage.

Stronach, I. and MacLure, M. (1997) *Educational research undone: The postmodern embrace*. Buckingham, UK: Open University Press.

Thatchenkery, T. (2004) Paradox and organisational change: The transformative power of hermeneutic appreciation. In D.L. Cooperrider and M. Avital (Eds.), *Constructive discourse and human organization: Advances in appreciative inquiry* (Vol. 1, pp. 77–104). Oxford, UK: Elsevier.

Trigwell, K. (2006) Phenomenography: An approach to research into geography education. *Journal of Geography in Higher Education*, 30(2), 367–372.

Turner, V. (1960) *The ritual process: Structure and anti-structure*. London: Routledge & Kegan Paul.

Usher, R., Bryant, I. and Johnstone, R. (1997) *Adult education and the postmodern challenge*. London: Routledge.

UK Evaluation Society (n.d.) Available online at: www.evaluation.org.uk/.

Vygotsky, L. (1962) *Thought and language*. Cambridge, MA: MIT Press.

Wagner, J. (2007) Observing culture and social life: Documentary photography, fieldwork and social research. In G.C. Stanczak (Ed.), *Visual research methods: Image, society and representation* (pp. 23–60). Los Angeles, CA: Sage.

Walford, G. (2001) *Doing qualitative educational research: A personal guide to the research process*. London: Continuum.

Webb, G. (1997) Deconstructing deep and surface: Towards a critique of phenomenography, *Higher Education*, 33(2): 195–212.

Weber, S. (2007) Analyzing visual qualitative data. In *Image Identity and Research Collective*. Montreal, Canada: Concordia University. Available online at http://iirc.mcgill.ca/txp/?sMethodology&c=About%20based%20 research.

Webster, L. and Mertova, P. (2007) *Using narrative inquiry as a research method: An introduction to using critical event narrative analysis in research on learning and teaching*. London: Routledge.

Wenger, E. (1998) *Communities of practice: Learning, meaning and identity*. Cambridge: Cambridge University Press.

Whyte, W.F. (1993) *Street Corner Society: The social structure of an Italian slum*. Chicago, IL: Chicago University Press.

Willis, P. (1977) *Learning to labour*. London: Saxon House.

Winter, R. (1998) Finding a voice – thinking with others: A conception of action research. *Educational Action Research*, 6(1), 53–68.

Wolff, K.H. (1995) *Transformation in the writing: A case of surrender-and-catch*. Dordrecht, Netherlands: Kluwer Academic.

Yin, R.K. (1993) *Applications of case study research*. Newbury Park, CA: Sage.

Yin, R.K. (2002) *Case study research, design and methods*, 3rd ed. Newbury Park, CA: Sage.

Zamorski, B. (2002) Research led teaching and learning in higher education. *Teaching in Higher Education*, 7(4), 411–427.

Zuber-Skerrit, O. (1996) *Action research in higher education: Examples and reflections*. London: Kogan Page.

Index

eBooks – at www.eBookstore.tandf.co.uk

A library at your fingertips!

eBooks are electronic versions of printed books. You can store them on your PC/laptop or browse them online.

They have advantages for anyone needing rapid access to a wide variety of published, copyright information.

eBooks can help your research by enabling you to bookmark chapters, annotate text and use instant searches to find specific words or phrases. Several eBook files would fit on even a small laptop or PDA.

NEW: Save money by eSubscribing: cheap, online access to any eBook for as long as you need it.

Annual subscription packages

We now offer special low-cost bulk subscriptions to packages of eBooks in certain subject areas. These are available to libraries or to individuals.

For more information please contact webmaster.ebooks@tandf.co.uk

We're continually developing the eBook concept, so keep up to date by visiting the website.

www.eBookstore.tandf.co.uk

LIBRARY, UNIVERSITY OF CHESTER